Springer Theses

Recognizing Outstanding Ph.D. Research

Aims and Scope

The series "Springer Theses" brings together a selection of the very best Ph.D. theses from around the world and across the physical sciences. Nominated and endorsed by two recognized specialists, each published volume has been selected for its scientific excellence and the high impact of its contents for the pertinent field of research. For greater accessibility to non-specialists, the published versions include an extended introduction, as well as a foreword by the student's supervisor explaining the special relevance of the work for the field. As a whole, the series will provide a valuable resource both for newcomers to the research fields described, and for other scientists seeking detailed background information on special questions. Finally, it provides an accredited documentation of the valuable contributions made by today's younger generation of scientists.

Theses may be nominated for publication in this series by heads of department at internationally leading universities or institutes and should fulfill all of the following criteria

- They must be written in good English.
- The topic should fall within the confines of Chemistry, Physics, Earth Sciences, Engineering and related interdisciplinary fields such as Materials, Nanoscience, Chemical Engineering, Complex Systems and Biophysics.
- The work reported in the thesis must represent a significant scientific advance.
- If the thesis includes previously published material, permission to reproduce this must be gained from the respective copyright holder (a maximum 30% of the thesis should be a verbatim reproduction from the author's previous publications).
- They must have been examined and passed during the 12 months prior to nomination.
- Each thesis should include a foreword by the supervisor outlining the significance of its content.
- The theses should have a clearly defined structure including an introduction accessible to new Ph.D. students and scientists not expert in the relevant field.

Indexed by zbMATH.

Chris Nagele

General Relativistic Instability Supernovae

Doctoral Theses accepted by
University of Tokyo, Tokyo, Japan

Springer

Author
Dr. Chris Nagele
Department of Astronomy
The University of Tokyo
Bunkyo City, Tokyo, Japan

Supervisor
Prof. Hideyuki Umeda
Department of Astronomy
School of Science
The University of Tokyo
Bunkyo City, Tokyo, Japan

ISSN 2190-5053   ISSN 2190-5061 (electronic)
Springer Theses
ISBN 978-981-96-0550-7   ISBN 978-981-96-0551-4 (eBook)
https://doi.org/10.1007/978-981-96-0551-4

© The Editor(s) (if applicable) and The Author(s), under exclusive license to Springer Nature Singapore Pte Ltd. 2024

This work is subject to copyright. All rights are solely and exclusively licensed by the Publisher, whether the whole or part of the material is concerned, specifically the rights of translation, reprinting, reuse of illustrations, recitation, broadcasting, reproduction on microfilms or in any other physical way, and transmission or information storage and retrieval, electronic adaptation, computer software, or by similar or dissimilar methodology now known or hereafter developed.
The use of general descriptive names, registered names, trademarks, service marks, etc. in this publication does not imply, even in the absence of a specific statement, that such names are exempt from the relevant protective laws and regulations and therefore free for general use.
The publisher, the authors and the editors are safe to assume that the advice and information in this book are believed to be true and accurate at the date of publication. Neither the publisher nor the authors or the editors give a warranty, expressed or implied, with respect to the material contained herein or for any errors or omissions that may have been made. The publisher remains neutral with regard to jurisdictional claims in published maps and institutional affiliations.

This Springer imprint is published by the registered company Springer Nature Singapore Pte Ltd.
The registered company address is: 152 Beach Road, #21-01/04 Gateway East, Singapore 189721, Singapore

If disposing of this product, please recycle the paper.

Supervisor's Foreword

Supermassive stars can be defined as stars which collapse by the general relativistic gravitational instability. Usual massive stars with the initial mass less than $\sim 140 M_\odot$ collapse by iron core collapse at the end of their evolution, while 140 300 $M_\odot$ stars collapse and explode (when mass loss is not strong) by the electron-positron pair instability. For these stars, general relativity is irrelevant to stellar collapse though its effects on dynamics are not small for iron core collapse.

Stars collapse by general relativistic effects only when their mass is very large as $10^5 \sim 10^6 M_\odot$. In these supermassive stars, radiation pressure is dominant, and the stars are marginally stable in Newtonian gravity, while they can be unstable in Einstein's gravity during the main-sequence stage. This fact was first discovered and applied to stars in 1960s. Since then, it has been occasionally proposed that supermassive stars were the central engine of QSOs, however, it is now widely believed that the engines are supermassive blackholes. Since the 2010's, observations of supermassive blackholes at high redshift AGNs unveiled severe problems. It appears that there is not enough time to form such massive blackholes from stellar mass blackhole seeds. One promising solution is to consider a very massive seed blackhole produced from a supermassive star, though there has been no observational evidence that such supermassive stars existed in the early universe. The coming decades will be an exciting era for these topics, since huge amounts of observational data about the early universe will be provided e.g., by the recently launched space telescope, JWST, several next generation telescopes, and gravitational wave observations.

To confront with these observations, it is necessary to polish predictions of a theoretical model. Regarding supermassive stars, by the end of 1980s it was considered that important properties of supermassive stars were already known, and the remaining problem was if and how these stars were formed. For example, it was considered that non-rotating Population III (metal-free) supermassive stars did not explode and collapse to a blackhole without mass ejection. However, in 2014 one counter example was discovered. There was a report that a Pop III 55,500 $M_\odot$ star exploded with huge explosion energy. This type of explosion was overlooked previously since the explosion occurs during the He-burning stage. Since they found only one explosion model, it was not clear if this discovery was robust or not. One of the main goals of Dr. Nagele's dissertation is to clarify this point.

In order to do this, a stability analysis code was newly developed to more precisely find the timing for general relativistic instabilities. Then the collapse/explosion simulations were performed using a general relativistic 1D hydrodynamical code with a detailed nuclear reaction network. With these procedures, this thesis could provide much more accurate answers for the fate of Pop III supermassive stars than before.

The results are interesting in two aspects. First, there certainly exist the mass ranges for explosion or mass ejection even for Pop III supermassive stars. Second, the mass ranges are large enough so that these explosion phenomena could be observed in near future. Such a discovery gives strong evidence for the existence of supermassive stars. In this thesis, the same methods are also applied to metal enriched and mass accreting cases. The results are all novel and will be exciting to verify in future observations.

Tokyo, Japan
August 2024

Hideyuki Umeda

Publications

Parts of this thesis have been published in the following journal articles

Nagele, C., Umeda, H., Takahashi, K., Yoshida, T., & Sumiyoshi, K. 2022, Monthly Notices of the Royal Astronomical Society, 517, 1584

Nagele, C., Umeda, H., Takahashi, K., & Maeda, K. 2023, Monthly Notices of the Royal Astronomical Society, 520, L72

Nagele, C., Umeda, H., & Takahashi, K. 2023, Monthly Notices of the Royal Astronomical Society 523, 1629

Nagele, C. & Umeda, H., The Astrophysical Journal, 949, L16 Nagele, C. & Umeda, H., Phys. Rev. D 110, L061301.

Acknowledgements

For his kindness and guidance, I would like to thank my adviser, Hideyuki Umeda. For their support and encouragement, across many miles and through challenging times, I would like to thank my parents, Robin and Drew Nagele, and my sister, Rose Nagele. For every hand who guided the course of this academic journey, Matt Kleban, Koh Takahashi, Takashi Yoshida, Kohsuke Sumiyoshi, Tilman Hartwig, John Silverman, Keiichi Maeda, and Takami Kuroda, I offer the hope that others will learn as much from you as I have. Finally, I must thank my friends in Japan, who have made my time here ever more unique, Yuta Nakagawa, Jake Butter, Walker Peterson, and Hitomi Terakawa.

Contents

1	**Introduction**		1
	1.1 Evidence for Supermassive Stars		2
		1.1.1 Early Universe Supermassive Black Holes	2
		1.1.2 Multiple Stellar Populations in Globular Clusters	4
	1.2 Formation of Supermassive Stars		4
		1.2.1 Atomic Cooling Halos	4
		1.2.2 Baryon-Dark Matter Supersonic Streaming	5
		1.2.3 Turbulent Cold Flows	5
		1.2.4 Gas Rich Galaxy Mergers	6
		1.2.5 Runaway Collisions in Nuclear Star Clusters	6
	1.3 Plausible Approaches to Detecting Supermassive Stars		7
	1.4 Non Degenerate Thermonuclear Supernovae		8
		1.4.1 Pair Instability Supernovae	8
		1.4.2 GR Instability Supernovae	9
	1.5 Novel Contributions of This Thesis		10
	References		11
2	**Methods**		15
	2.1 Stellar Evolution		16
		2.1.1 Initial Conditions	18
		2.1.2 Mass Loss	19
		2.1.3 Linearized Equation of Motion in GR	19
		2.1.4 Necessary Condition for Stability	23
		2.1.5 Evaluation of This Condition on Numerical Models	24
		2.1.6 Accretion	26
	2.2 Hydrodynamics		27
		2.2.1 Numerical Scheme	27
		2.2.2 Nuclear Networks	30
		2.2.3 Consistency with Evolutionary Profiles and Stability Analysis	31

	2.3	Supernova Lightcurves	33
		2.3.1 SNEC	33
	2.4	Supernova Rate Estimation	34
		2.4.1 Metal Free	34
		2.4.2 Metal Rich	34
	References		34
3	**Results**		37
	3.1	Stellar Evolution	37
		3.1.1 GR Stability	37
		3.1.2 Metal Free	40
		3.1.3 Metal Rich	41
		3.1.4 Accreting	42
	3.2	Hydrodynamics	45
		3.2.1 Metal Free	46
		3.2.2 Metal Rich	52
		3.2.3 Accreting	55
	3.3	Supernova Lightcurves	57
		3.3.1 Metal Free	57
	3.4	Nucleosynthetic Yields	59
		3.4.1 Metal Free	59
		3.4.2 Metal Rich	59
	3.5	Comparison to Super-Solar Nitrogen in GN-Z11	62
	3.6	Prospects Using Current and Future Observatories	65
	References		66
4	**Discussion**		67
	References		69
5	**Conclusion**		71

Chapter 1
Introduction

Abstract Two of the most challenging open problems in astronomy are the origin of the early universe supermassive black holes and the origin of multiple stellar populations in globular clusters. Both of these problems have been suggested to have solutions involving supermassive stars, which are stars that experience the general relativistic radial instability. We begin with a review of the history of the study of supermassive stars, and the motivation for studying them. Supermassive stars, however, are most likely to exist in the high redshift universe, and therefore they would be difficult to detect, though promising avenues such as strong lensing and gravitational waves from black hole formation are being pursued. In this thesis, we focus on a different method of detection, namely thermonuclear explosions occurring at the end of the lives of supermassive stars, events which have been termed general relativistic instability supernovae.

Keywords Supermassive black hole problem · Early universe · Thermonuclear supernovae

Astronomy is an astonishingly successful field of study. In the last decade alone, astronomers have discovered gravitational waves from coalescing compact objects, as well as a coincident kilonova, a stochastic gravitational wave background, a new order of magnitude of exo-planets, solar neutrinos, fast radio bursts, and this was all before the James Webb Space Telescope (JWST) scythed through existing redshift limits.

Given, however, the reasonable supposition that humanity will not be leaving the solar system any time soon, the question of why we study the myriad objects beyond the sun's influence should be addressed. There are, of course, a multitude of satisfying answers to this question, but the one that resonates with me is the ability of astronomers to test fundamental physics, such as nuclear reaction rates, general relativity, inflationary cosmology, and dense matter. Simply put, if physics presents a theory of the universe, then astronomy should determine if that theory can explain the canon of observation. If it can, then the theory may be valid. If not, then physics has more work to do.

Yet despite its recent successes, astronomy cannot fully hold up its end of the bargain. Observationally challenging ideas from physics such as the existence of primordial black holes cannot be ruled out because they solve problems to which astronomy has no agreed upon solution, such as the nature of dark matter or the origin of early universe supermassive black holes. In this context, it is clear that advances in astronomy are critical to formulating a coherent theory of physics.

1.1 Evidence for Supermassive Stars

A supermassive star (SMS) is a star which ends its life due to the general relativistic radial instability (GR instability, [13]), and fulfilling this condition requires these stars to typically be larger than 10^4 $M_\odot$. They were first conceived as a means of explaining the large radio emission of quasi-stellar objects (QSOs) [37], and after QSOs were shown to arise from accreting supermassive black holes (SMBHs) it was suggested that a smaller, variable, population of the QSOs (type I Seyfert galaxies) could be explained by SMSs [90]. This idea also did not gain traction, and the main interest in SMSs today derives from their possible contribution to the *formation* of SMBHs, as was suggested by [8], although see [80] for a more fleshed out discussion. The evolution and stability of SMSs has been studied for some time [4, 23, 24, 38], but this study has been made much more robust with the use of stellar evolution and hydrodynamics codes (e.g. [14, 25, 30, 35, 36, 99, 112]) which allow greater confidence in predicting the proprieties of SMSs. As these object have yet to be observed, however, we now turn to the evidence for their existence.

1.1.1 Early Universe Supermassive Black Holes

SMBHs are observed soon after the big bang [5, 20, 21, 55, 56, 64, 107, 113], and this is a huge problem for astronomy (for recent reviews, see e.g. [40, 103, 106]). Our current understanding of cosmology requires that the SMBHs did not exist at the time of the big bang, implying that they were created in the brief intervening period. The simplest proposal for creating a SMBH is to take a $\sim$30 $M_\odot$ black hole, which we know exist in the local universe from gravitational wave observations [1] and are also theorized to exist in the high redshift universe (e.g. [48]), and accrete large amounts of matter onto it in the gas rich early universe. However, this proposal is severely hampered by the Eddington limit (e.g. [95]), meaning that even if the $\sim$30 $M_\odot$ black hole were to grow continuously at the Eddington rate (which is itself an unrealistic assumption), it would not reach the observed quasar masses at the observed cosmic times [40].

How can we approach solving this seemingly unsolvable problem? The most common solution at current is to invoke the presence of a massive seed black hole which can grow to the required masses without requiring super-Eddington accretion. This almost always requires the formation of a SMS which then collapses into this seed black hole, although notable exceptions are discussed below. The other main

1.1 Evidence for Supermassive Stars

approach is to suggest that the black hole can grow faster than the Eddington limit, either through super-Eddington accretion or through BH mergers. The rest of this subsection will discuss these two suggestions in detail.

The main topic of this thesis is the SMSs which collapse to massive BH seeds, but do other pathways to constructing massive BH seeds exist? Part of this discussion descends into the semantic. We define a SMS as an object which is hydrostatic on nuclear timescales (e.g. [49]), so that if a supermassive protostar which has not yet begun nuclear burning collapses, we do not classify it as a SMS. Such collapse is possible in the galaxy merger scenario where extremely large accretion rates onto a protostar of up to 10^5 $M_\odot$/year are expected [59, 67]. The resulting black hole is expected to be more massive than 10^5 $M_\odot$. Another case which avoids the classification of SMS is the dark star scenario (e.g. [6]), in which 10^5 $M_\odot$ protostars are formed, but are supported by dark matter annihilation instead of nuclear burning. In some cases, these dark stars will transition to SMSs, but if they grow to be massive enough, then they are expected to collapse directly to back holes. Finally, relatively smaller black hole seeds ($M \sim 10^3$ $M_\odot$) are expected to arise from very massive stars (those that end their lives due to the pair instability instead of the GR instability) produced by runaway collisions in nuclear star clusters. However, as some of these objects are expected to be more massive SMSs, we postpone their discussion to Sect. 1.2.5.

Next, we turn to the possibility of primordial black holes providing the massive seeds (e.g. [106]). Primordial black holes are black holes which are formed by matter over-densities in the early universe, where the mass of the PBH corresponds to the horizon size at the time of formation [10]. This means that PBHs could feasibly have masses as small as 10^{-5} g and as large as 10^5 $M_\odot$. Although robust constraints exist against PBHs making up dark matter at a single mass, if PBHs only contribute a slight fraction of DM, or if they span a range of masses, then ruling them out observationally may be difficult. Thus, PBHs can feasibly solve the early universe supermassive black hole problem, but confirming this solution may be more challenging than in the other scenarios.

Another possibility is that stellar mass black holes accreted material more rapidly than the Eddington limit allows. Although most accretion onto SMBHs is thought to occur well below the Eddington limit, some quasars and x-ray binaries are accreting at or above the Eddington limit. In most cases, there is thought to be a duty cycle, ultimately driven by radiative feedback, so that the time averaged accretion rate does not exceed the Eddington limit (e.g. [40]). However, there does exist another regime, specifically hyper-Eddington accretion where photons which would normally induce radiative feedback are instead trapped by the inflow (e.g. [39]). This phenomenon has been explored in 1D calculations [39, 44, 83, 114] and multiple dimensions [43, 73, 89], although the precise conditions for hyper-Eddington accretion are still being explored (see discussion in e.g. [44]) and thus it is still unknown whether this is a viable mechanism towards resolving the early universe SMBH problem.

Finally, we examine the contribution from black hole mergers. Galaxies in the early universe have rich merger histories, and if each galaxy merger eventually leads to a SMBH merger, then this process could meaningfully contribute to SMBH growth

[106]. Evidence for binary SMBH systems include the observation of dual AGN (e.g. [53]) and the recent detection of a stochastic gravitational wave background by NANOgrav [2]. However, direct evidence of mergers themselves (and a measurement of the merger rate) will likely have to wait for LISA [3].

In summary, supermassive stars are by no means the only solution to the early universe SMBH problem; however, the community suspects that massive seeds will be one part of this solution [40] and SMSs are likely the easiest of the massive seed scenarios to detect and to model, due to their similarity to massive stars for which extensive modeling has been already carried out.

1.1.2 Multiple Stellar Populations in Globular Clusters

Another, completely disparate, collection of evidence also suggests the existence of SMSs. This aims to explain the C–N and O–Na abundance anti-correlations in globular clusters [11, 12] which can be resolved by invoking hot CNO nucleosynthesis [78]. The conditions for this nucleosynthesis can be reached in a very/supermassive star (V/SMS) formed by runaway collisions [77] and this scenario can explain the anti-correlations if enrichment ceases before much helium is produced [17, 18]. Furthermore, the observation of multiple stellar populations [76] is consistent with a V/SMS polluter if the star formation occurs during the lifetime of the V/SMS [28].

1.2 Formation of Supermassive Stars

In this section, we outline the various formation mechanisms of SMSs. The scenarios outlined in Sects. 1.2.1–1.2.3 are thought to occur only in environments with very low metallicity [16, 34], while the scenarios in Sects. 1.2.4 and 1.2.5 do not exclude higher metallicity environments, although there may exist a metallicity dependence.

1.2.1 Atomic Cooling Halos

The idea behind atomic cooling halos is a relatively simple one. Consider that the Jeans mass, often equated with the final mass of a star, is proportional to $T^{3/2}$. Thus, if the gas that eventually becomes a star is hotter, then that star will be more massive. Throughout most of cosmic time, there have been sufficient metals in ambient gas so that this gas can cool to very low temperatures ($\sim$10–100 K) by photons emitted via electron transitions in metals. In the early universe, however, no such metals, and no such cooling mechanism, exist. Still, in most cases, molecular hydrogen can cool gas via the vibrational mode, but if for some reason these molecules are removed, then the gas cannot cool below 8000 K. These instances are referred to as atomic cooling (as opposed to molecular cooling) halos (short for dark matter halos, the first virialized objects).

How might molecular hydrogen be destroyed? The most common scenario in the literature is that there is a local flux of high energy Lyman Werner photons which photodissociates the molecules [9, 72, 74]. This flux could arise from massive Pop III stars or supernovae, so this scenario requires two nearby star forming regions close enough so that photons in the first can photodissociate molecular hydrogen in the second, but far enough away so that metal enrichment from the first cannot pollute the pristine gas of the second. This effect can be enhanced by collisional dissociation [22] if there is sufficient turbulence within the halo. Alternatively, if the halo undergoes frequent mergers, then dynamical heating can help to maintain the high temperature of the gas [111]. Finally, if there is a head on collision of two such halos at relatively high velocity, then both dynamical heating and collisional photodissociation of H_2 are thought to occur, thus obviating the need for a Lyman Werner flux [41].

1.2.2 Baryon-Dark Matter Supersonic Streaming

Another popular method of forming SMSs is by the effect of baryon-dark matter supersonic streaming [96]. Throughout the universe, baryonic matter and dark matter have relative motions [98, 102]. However, in some extreme cases, that relative motion can be high enough that it suppresses normal star formation. In this case, the dark matter halo continue to grow through accretion and mergers, while the baryonic matter is prevented from forming stars [33, 50, 84, 95]. Eventually, the gravitational well of the halo becomes so massive that the local turbulence is overcome and gravitational collapse of the baryonic matter occurs. However, at this time, so much material has accumulated that the resulting star is supermassive, even if H_2 cools the gas [33].

1.2.3 Turbulent Cold Flows

A relatively recent proposal for producing SMSs is via the use of turbulent cold flows [51]. These are accretion flows which appear in cosmological simulations and this proposal was originally motivated by the coincidence of the accretion flows with the high redshift (z) quasars in these simulations [51, 97]. The basic idea is that these flows collide with each other and with pre-exiting gas in the halo, and those collisions induce local turbulence which suppresses star formation, similar to the way that the baryonic-dark matter supersonic streaming suppressed star formation in the previous subsection. Since star formation is suppressed, the halo continues to grow, and eventually gravitational collapse occurs, but by that time the mass is so large that SMSs form [51]. It should be noted that these cold flows are thus responsible both for the initial SMBH seed and for the subsequent accretion up to the masses of observed quasars. This coincidence means that the turbulent cold flow model would lose some attractiveness if the cosmological simulations were to be modified (either structurally, or in the accretion prescriptions), but nevertheless, the cold flow model is extremely promising.

1.2.4 Gas Rich Galaxy Mergers

In the galaxy merger scenario, SMS formation is triggered by gravitational torques during the merger of two gas rich galaxies (e.g. [57]). The phenomenon of nuclear gaseous disks forming via multi-scale inflows was first investigated in the context of the M-sigma relation as a means of providing a source of dynamical friction for a SMBH binary in order to assist with its eventual merger [47, 60]. Since then, it has been shown that not only can this disk influence the behavior of existing SMBHs, but it can also collapse under its own gravity to form a new black hole [59]. Confirmation of this scenario may be possible with LISA if the central object maintains sufficient asphericity inherited from the nuclear disk [115]. An enticing ingredient of this scenario is that it is agnostic to the metallicity of the interstellar medium (ISM), meaning SMSs will form out of metal enriched gas [58].

The galaxy merger scenario predicts extremely high infall rates (up to $\sim 10^4$ $M_\odot$/year) over relatively short periods of time ($\sim 10^4$ years) on parsec scales. These rapid inflows then translate to possibly comparable accretion rates (see [58]) onto a supermassive protostar. In the most extreme cases, specifically when both merging galaxies have virial masses greater than 10^{11} $M_\odot$, the protostar will continue to accrete gas until it collapses due to the GR radial instability. However, in the mergers of less massive galaxies, infall rates of order 10^3 $M_\odot$/year are expected. The protostars formed by these inflows eventually transition to supermassive stars (M $\sim 10^{3-5}$ $M_\odot$) due to reservoir exhaustion [58, 82, 85].

1.2.5 Runaway Collisions in Nuclear Star Clusters

Finally, we turn to the phenomenon of runaway collisions in nuclear star clusters [77]. Note that the community usually distinguishes between direct collapse black holes (DCBH, Sects. 1.2.1–1.2.4) and nuclear star clusters (NSC, this section) by claiming that they leave behind seeds of roughly 10^5 and 10^3 $M_\odot$ respectively. We will not make this distinction because it is likely that there exists a large amount of overlap between the seed masses of DCBH and NSC, especially if the GR instability triggers (e.g. [28]).

The usual requirement for runaway collisions is that a star cluster is accreting gas which does not significantly alter the angular momentum of the cluster. Then, the increased mass leads to gravitational contraction and possible stellar collisions

[61]. Once a few collisions have occurred, a V/SMS forms and the rate of collisions accelerates (hence runaway) due to the huge radius of the V/SMS [28]. As with Sect. 1.2.4, a SMS formed in this manner can be metal rich. Thus line driven mass loss is important, and the evolutionary history of the star becomes a competition between accretion and mass loss. These stars are hypothesized to be fully convective so that mass lost to winds will contain hot CNO nucleosynthetic material produced in the center of the star. It is this material which may explain the anti-correlations in globular clusters and the presence of multiple stellar populations (Sect. 1.1.2).

1.3 Plausible Approaches to Detecting Supermassive Stars

As discussed above, the discovery of SMSs would shed light on the mass of the seeds for early universe SMBHs as well as aid in our understanding of the origin of multiple stellar populations in globular clusters. SMSs have not been detected so far, however, and are thought to live mostly in the early universe, so is there reasonable hope for finding them?

The first and most obvious approach is to directly search for the SMS photosphere, possibly aided by strong lensing [92, 93, 104]. Surace et al. [93] calculated the spectra of red SMSs surrounded by an accretion envelope. These SMSs at high redshift ($z \approx 20$) are ideal for detection by JWST and other IR instruments because the low effective temperature means most of the photons are redward of the Lyman break. On the other hand, blue SMSs [92] are still detectable by EUCLID and JWST with modest strong lensing, but only out to redshifts of about $z \approx 10$. Vikaeus et al. [104] extended this study to include Roman, which has a wider field of view than JWST and deeper sensitivity than EUCLID.

Other, more esoteric, methods of detecting SMSs have also been suggested. Ultra-long gamma ray bursts (UL-GRBs) are a class of GRBs which last for thousands of seconds, much longer than the typical duration of about 10 s for long GRBs. There are several models which could feasibly explain UL-GRBs ([75], references therein) , and one of these is the collapse of a rapidly rotating SMS to a black hole [91]. The black hole formation is expected to launch a jet just as less massive stars are though to when forming black holes, but the accretion disk (or torus) which accretes onto the black hole contains a huge amount of material and is thus long lived (10^5 s), prompting the suggestion that this event could be associated with a UL-GRB.

Sun et al. [91] also calculated the gravitational wave signal from black hole formation and ringdown. Similar calculations were also carried out by [88]. Both studies found that these events would be visible to LISA [3] at redshift 3 and [91] also found that they would be visible to DECIGO [46] at redshift 5. Note, however, that both of these studies used rapidly rotating polytropic progenitors, and the assumption of rapid rotation may conflict with evolutionary models of SMSs due to the Omega-Gamma limit [31, 54].

Finally, the neutrino signal from SMS collapse is extremely large, reaching luminosities of $\sim 10^{56}$ ergs/s [52, 65, 70, 87]. This neutrino flux could be feasibly detected from a nearby (1 Mpc) SMS collapse [65, 70], but other pathways are more challenging. [87] suggested that SMS collapse could meaningfully contribute to the relic neutrino background, but [70] concluded that this was unrealistic. Finally, [52] suggested that neutrino emission could produce a linear memory gravitational wave burst ([81], references therein) detectable by DECIGO, but they used the results of [87] to do so, which may overestimate the neutrino luminosity.

1.4 Non Degenerate Thermonuclear Supernovae

Perhaps the most promising avenue toward detection, however, is a GRSN, and we now turn to the topic of supernovae powered by explosive nuclear burning. While we discuss two different types of thermonuclear supernovae in this section, we first lay out some general considerations.

Non degenerate thermonuclear supernovae (so called to distinguish them from degenerate SNIa) occur when the evolution of a massive star towards forming an iron core is disrupted by some sort of instability. After this instability, the star contracts and the temperature and density leading to explosive nuclear burning. Because the star has not yet formed an iron core, these reactions produce energy, thus increasing the overall energy of the star (although neutrino reactions slightly counteract this trend). If the energy of the star is increased to such an extent that the total energy is larger than the kinetic energy of the infalling matter, then a shock forms at the center of the star which propagates outwards, triggering an explosion and a supernova.

The amount of nuclear burning (e.g. increase in energy density) which takes place during the explosion is greatest in the center of the star, and decreases smoothly towards the outer regions of the star, where lower temperatures are reached. Thus the maximum central temperature is a useful diagnostic for the strength of the explosion. For each of the below cases, very weak explosions do not eject the entire model, but instead eject only a portion of the envelope and these events are known as pulsations, as they are thought to be able to occur multiple times [68].

1.4.1 Pair Instability Supernovae

A pair-instability supernova (PISN) is a thermonuclear explosion of a massive Oxygen core induced by the creation of electron-positron pairs. This mechanism was originally proposed by [7, 79] and eventually confirmed through numerical modeling (e.g. [32, 94, 101]). In order for a star to experience the pair instability, it must be very massive (~ 100–$1000\,M_\odot$) during the oxygen burning phase. Because metal poor stars are thought to have a more top heavy IMF [15] and because of the strong dependence of mass loss rates with metallicity (e.g. [105]), PISN are thought to prefer lower metallicities.

1.4 Non Degenerate Thermonuclear Supernovae

The pair instability triggers during carbon or oxygen burning. Typical maximum temperatures reached are Log T_c = 9.5–9.8. If the central temperature becomes high enough, then large quantities of ^{56}Ni are produced. This is not necessarily the termination of the α process, as in core collapse SNe (CCSNe), but rather because of the low densities in the star, protons are freely available and reactions such as proton captures and inverse beta decays allow lighter isotopes to reach intermediate masses (for instance, iron peak elements). The consequent supernova is thought to experience two phases, shock breakout and cooling, followed by a nickel dominated phase.

PISNe have yet to be definitively detected. The first suggestion of a PISN detection was reported as a Type I super luminous supernova (SLSN-I), SN2007bi [27]. However, spectroscopic models of PISNe were incompatible with the observations both in the photospheric [19] and nebular [42] phases. Since then, magnetar models have more frequently been considered to explain SLSNe-I, and PISNe models are rarely taken seriously. Nevertheless there remain some SLSNe-I, such as PS1-14bj, PTF10nmn, OGLE14, and SN2020wnt, which cannot be well explained by the magnetar models and these could be PISNe. If these candidates are PISNe, the ejected ^{56}Ni mass falls in the range of $\sim$0.5–10 $M_\odot$ [26]. Very recently, there was a report that SN2018ibb might be the best PISN candidate discovered thus far [86]. SN2018ibb is a Hydrogen poor super-luminous SN at $z = 0.166$, and if it is a PISN, the expected ^{56}Ni mass is more than 30$M_\odot$. [86] also mention that there is a signature of interaction with oxygen-rich CSM and models fitting both of these conditions were recently reproduced for rotating sub-solar metallicity progenitors [100].

1.4.2 GR Instability Supernovae

GRSNe are similar to PISNe except that they are triggered by the GR radial instability (Sec. exrefsec:methodsspsGRstabspsnumerical) instead of the pair instability. This requires even more massive progenitor stars (this is, in fact, the definition of a SMS). There are two broad categories of GRSNe, those which occur during helium burning and those which occur during hydrogen burning.

Helium burning GRSNe were first discovered by [14], who performed a grid of post-Newtonian (PN) stellar evolution simulations for SMSs of varying mass using the KEPLER code. They found that the GR instability triggering during late helium burning could cause an incredibly energetic explosion ($E_{\exp} \approx 6$–9×10^{54} ergs) for one specific model with mass 55500 $M_\odot$. The explosion starts with the ^{16}O(α, γ) reaction and then proceeds along the α process before terminating at ^{28}Si and ^{32}S. Subsequent work then investigated some of the consequences that such an energetic event would have [45, 108, 109]. Recently, we confirmed that an explosion around this mass range was possible, although our explosion energy was much lower and we found that a specific envelope structure was required for the explosion [69]. However, both of these studies suffered from a failure to accurately evaluate the GR radial instability. In subsequent work, we showed that such an evaluation leads to a much more robust phenomenon [71], and that is a large part of the work presented in this thesis.

Hydrogen burning GRSNe occur for metal enriched SMS progenitors which experience the GR instability during the hydrogen burning phase. They were first investigated by [25] who used the KEPLER code to show that models around 10^5 $M_\odot$ would explode with sufficient metallicity and collapse to BHs otherwise. [62] conducted a similar study using a 2D BSSN code with parameterized heating rates, and found similar results for their non-rotating models. The reason that enhanced metallicity is required for an explosion is that the p-p chain is too slow to explode the star, but the CNO cycle is fast enough. Thus, the presence of carbon allows an explosion to take place. Also of importance, however, is the presence of other metals with serve as seeds for proton captures, thus enabling the rapid proton capture process (rp-process) as we will show. The hydrogen burning GRSNe have even larger explosion energies ($E_{\mathrm{exp}} \sim 10^{55-56}$ ergs), due to more massive progenitors and the fact that hydrogen burning releases more energy per unit mass.

Both types of GRSN are extremely energetic and have characteristic yields. Thus, searching for the lightcurves associated with these explosions or their distinct nucleosynthetic patterns are extremely promising pathways towards confirming the existence of SMSs, and thus alleviating some of the difficulties discussed above.

1.5 Novel Contributions of This Thesis

The primary advancement of the study of GRSNe shown in this thesis is the GR stability analysis [71]. This analysis discovered that the GR radial instability occurs earlier than was previously thought [25, 29]. It also revealed that the stability of SMSs can depend on the envelope structure of the star, behavior that is stochastic with respect to the total mass of the star. The combination of these two facts means that GRSNe are much more common, spanning wider ranges of mass and metallicity, than was previously thought [14, 25, 62].

Another consequence of the use of the GR stability analysis was the discovery of pulsations, thermonuclear events which eject mass but do not completely unbind the star [68]. These pulsations occur in the late helium burning phase in metal free SMSs, which have very large radii. Since luminosity scales roughly with radius, the associated lightcurves of these events are even brighter than the explosions (see also [63, 110]). We were also able to show that these pulsations can occur more than once.

The final novel ingredient in this thesis is the modeling of the hydrodynamical phase of the explosion using a fully relativistic hydrodynamics code coupled to large nuclear networks [67]. Except for [62], previous works had only used post Newtonian gravity [14, 25] which, sometimes severely, underestimates the gravitational force. Furthermore, previous works used small nuclear networks [14, 25] or parameterized heating rates [62]. Our use of large nuclear networks permits more confidence in our results and allows us to compute detailed nucleosynthetic yields. Indeed, the comparison of these yields to JWST observations of GN-z11 is a major component of this thesis [66].

References

1. Abbott R, Abbott TD, Acernese F et al (2023) Phys Rev X 13:011048
2. Agazie G, Anumarlapudi A, Archibald AM et al (2023). arXiv e-prints arXiv:2306.16220
3. Amaro-Seoane P, Audley H, Babak S et al (2017). arXiv e-prints arXiv:1702.00786
4. Appenzeller I, Fricke K (1972) Astron Astrophys 21:285
5. Bañados E, Venemans BP, Mazzucchelli C et al (2018) Nature 553:473
6. Banik N, Tan JC, Monaco P (2019) Mon Not R Astron Soc 483:3592
7. Barkat Z, Rakavy G, Sack N (1967) Phys Rev Lett 18:379
8. Begelman MC, Rees MJ (1978) Mon Not R Astron Soc 185:847
9. Bromm V, Loeb A (2003) Astrophys J 596:34
10. Carr B, Kühnel F (2020) Annu Rev Nucl Part Sci 70:355
11. Carretta E, Bragaglia A, Gratton R, Lucatello S (2009) Astron Astrophys 505:139
12. Carretta E, Bragaglia A, Gratton RG et al (2009) Astron Astrophys 505:117
13. Chandrasekhar S (1964) Astrophys J 140:417
14. Chen K-J, Heger A, Woosley S et al (2014) Astrophys J 790:162
15. Chon S, Hosokawa T, Omukai K (2021) Mon Not R Astron Soc 502:700
16. Chon S, Omukai K (2020) Mon Not R Astron Soc 494:2851
17. Denissenkov PA, Hartwick FDA (2014) Mon Not Roy Astron Soc 437:L21
18. Denissenkov PA, VandenBerg DA, Hartwick FDA et al (2015) Mon Not Roy Astron Soc 448:3314
19. Dessart L, Hillier DJ, Waldman R, Livne E, Blondin S (2012) Mon Not Roy Astron Soc 426:L76
20. Eilers A-C, Simcoe R A, Yue M et al (2022). arXiv e-prints arXiv:2211.16261
21. Fan X, Banados E, Simcoe RA (2022). arXiv e-prints arXiv:2212.06907
22. Fernandez R, Bryan GL, Haiman Z, Li M (2014) Mon Not R Astron Soc 439:3798
23. Fricke KJ (1973) Astrophys J 183:941
24. Fricke KJ (1974) Astrophys J 189:535
25. Fuller GM, Woosley SE, Weaver TA (1986) Astrophys J 307:675
26. Gal-Yam A (2019) Annu Rev Astron Astrophys 57:305
27. Gal-Yam A, Mazzali P, Ofek EO et al (2009) Nature 462:624
28. Gieles M, Charbonnel C, Krause MGH et al (2018) Mon Not R Astron Soc 478:2461
29. Haemmerlé L (2021) Astron Astrophys 647:A83
30. Haemmerlé L, Meynet G, Mayer L et al (2019) Astron Astrophys 632:L2
31. Haemmerlé L, Woods TE, Klessen RS, Heger A, Whalen DJ (2018). Astrophys J l 853:L3
32. Heger A, Woosley SE (2002) Astrophys J 567:532
33. Hirano S, Hosokawa T, Yoshida N, Kuiper R (2017) Science 357:1375
34. Hirano S, Machida MN, Basu S (2022). arXiv e-prints arXiv:2209.03574
35. Hosokawa T, Omukai K, Yorke HW (2012) Astrophys J 756:93
36. Hosokawa T, Yorke HW, Inayoshi K, Omukai K, Yoshida N (2013) Astrophys J 778:178
37. Hoyle F, Fowler WA (1963) Mon Not R Astron Soc 125:169
38. Iben, Icko J (1963) Astrophys J 138:1090
39. Inayoshi K, Haiman Z, Ostriker JP (2016) Mon Not R Astron Soc 459:3738
40. Inayoshi K, Visbal E, Haiman Z (2020) Annu Rev Astron Astrophys 58:27
41. Inayoshi K, Visbal E, Kashiyama K (2015) Mon Not R Astron Soc 453:1692
42. Jerkstrand A, Smartt SJ, Inserra C et al (2017) Astrophys J 835:13
43. Jiang Y-F, Stone JM, Davis SW (2019) Astrophys J 880:67
44. Johnson JL, Upton Sanderbeck PR (2022) Astrophys J 934:58
45. Johnson JL, Whalen DJ, Even W et al (2013) Astrophys J 775:107

46. Kawamura S, Ando M, Seto N et al (2011) Class Quant Gravity 28:094011
47. Kazantzidis S, Mayer L, Colpi M et al (2005) Astrophys J Lett 623:L67
48. Kinugawa T, Nakamura T, Nakano H (2020) Mon Not R Astron Soc 498:3946
49. Kippenhahn R, Weigert A, Weiss A (2012). Stellar Struct Evol. https://doi.org/10.1007/978-3-642-30304-3
50. Latif MA, Niemeyer JC, Schleicher DRG (2014) Mon Not R Astron Soc 440:2969
51. Latif MA, Whalen DJ, Khochfar S, Herrington NP, Woods TE (2022) Nature 607:48
52. Li J-T, Fuller GM, Kishimoto CT (2018) Phys Rev D 98:023002
53. Liu X, Civano F, Shen Y et al (2013) Astrophys J 762:110
54. Maeder A, Meynet G (2000) Astron Astrophys 361:159
55. Marshall MA, Perna M, Willott CJ et al (2023). arXiv e-prints arXiv:2302.04795
56. Matsuoka Y, Onoue M, Kashikawa N et al (2019) Astrophys J Lett 872:L2
57. Mayer L, Bonoli S (2019) Rep Progress Phys 82:016901
58. Mayer L, Fiacconi D, Bonoli S et al (2015) Astrophys J 810:51
59. Mayer L, Kazantzidis S, Escala A, Callegari S (2010) Nature 466:1082
60. Mayer L, Kazantzidis S, Madau P et al (2007) Science 316:1874
61. Moeckel N, Clarke CJ (2011) Mon Not R Astron Soc 410:2799
62. Montero PJ, Janka H-T, Müller E (2012) Astrophys J 749:37
63. Moriya TJ, Chen K-J, Nakajima K, Tominaga N, Blinnikov SI (2021) Mon Not R Astron Soc 503:1206
64. Mortlock DJ, Warren SJ, Venemans BP et al (2011) Nature 474:616
65. Munoz V, Takhistov V, Witte SJ, Fuller GM (2021). arXiv e-prints arXiv:2102.00885
66. Nagele C, Umeda H (2023) Astrophys J Lett 949:L16
67. Nagele C, Umeda H, Takahashi K (2023a). arXiv e-prints arXiv:2301.01941
68. Nagele C, Umeda H, Takahashi K, Maeda K (2023) Mon Not R Astron Soc 520:L72
69. Nagele C, Umeda H, Takahashi K, Yoshida T, Sumiyoshi K (2020) Mon Not R Astron Soc 496:1224
70. Nagele C, Umeda H, Takahashi K, Yoshida T, Sumiyoshi K (2021) Mon Not R Astron Soc 508:828
71. Nagele C, Umeda H, Takahashi K, Yoshida T, Sumiyoshi K (2022) Mon Not R Astron Soc 517:1584
72. Oh SP, Haiman Z (2002) Astrophys J 569:558
73. Ohsuga K, Mori M, Nakamoto T, Mineshige S (2005) Astrophys J 628:368
74. Omukai K (2001) Astrophys J 546:635
75. Perna R, Lazzati D, Cantiello M (2018) Astrophys J 859:48
76. Piotto G, Milone AP, Bedin LR et al (2015) Astron J 149:91
77. Portegies Zwart SF, Makino J, McMillan SLW, Hut P (1999) Astron Astrophys 348:117
78. Prantzos N, Charbonnel C, Iliadis C (2017) Astron Astrophys 608:A28
79. Rakavy G, Shaviv G (1967) Astrophys J 148:803
80. Rees MJ (1984) Annu Rev Astron Astrophys 22:471
81. Richardson CJ, Zanolin M, Andresen H et al (2022) Phys Rev D 105:103008
82. Sakurai Y, Hosokawa T, Yoshida N, Yorke HW (2015) Mon Not R Astron Soc 452:755
83. Sakurai Y, Inayoshi K, Haiman Z (2016) Mon Not R Astron Soc 461:4496
84. Schauer ATP, Regan J, Glover SCO, Klessen RS (2017) Mon Not R Astron Soc 471:4878
85. Schleicher DRG, Palla F, Ferrara A, Galli D, Latif M (2013) Astron Astrophys 558:A59
86. Schulze S, Fransson C, Kozyreva A et al (2023). arXiv e-prints arXiv:2305.05796
87. Shi X, Fuller GM (1998) Astrophys J 503:307
88. Shibata M, Sekiguchi Y, Uchida H, Umeda H (2016) Phys Rev D 94:021501
89. Sądowski A, Narayan R (2016) Mon Not R Astron Soc 456:3929
90. Stoner R, Ptak R (1984) Astrophys J 280:516
91. Sun L, Paschalidis V, Ruiz M, Shapiro SL (2017) Phys Rev D 96:043006
92. Surace M, Zackrisson E, Whalen DJ et al (2019) Mon Not Roy Astron Soc 488:3995
93. Surace M, Whalen DJ, Hartwig T et al (2018) Astrophys J Lett 869:L39
94. Takahashi K, Yoshida T, Umeda H (2018) Astrophys J 857:111

References

95. Tanaka T, Haiman Z (2009) Astrophys J 696:1798
96. Tanaka TL, Li M, Haiman Z (2013) Mon Not R Astron Soc 435:3559
97. Tenneti A, Di Matteo T, Croft R, Garcia T, Feng Y (2018) Mon Not R Astron Soc 474:597
98. Tseliakhovich D, Hirata C (2010) Phys Rev D 82:083520
99. Umeda H, Hosokawa T, Omukai K, Yoshida N (2016) Astrophys J Lett 830:L34
100. Umeda H, Nagele C (2023). arXiv e-prints arXiv:2307.02692
101. Umeda H, Nomoto K (2002) Astrophys J 565:385
102. Uysal B, Hartwig T (2023) Mon Not R Astron Soc 520:3229
103. Valiante R, Agarwal B, Habouzit M, Pezzulli E (2017) On the formation of the first quasars. Publ Astron Soc Aust 34:e031
104. Vikaeus A, Whalen DJ, Zackrisson E (2022). arXiv e-prints arXiv:2205.14163
105. Vink JS, de Koter A, Lamers HJGLM (2001) Astron Astrophys 369:574
106. Volonteri M, Habouzit M, Colpi M (2021) Nat Rev Phys 3:732
107. Wang F, Yang J, Fan X et al (2021) Astrophy J Lett 907:L1
108. Whalen DJ, Johnson JL, Smidt J et al (2013) Astrophys J 777:99
109. Whalen DJ, Johnson JL, Smidt J et al (2013) Astrophys J 774:64
110. Whalen DJ, Even W, Smidt J et al (2013) Astrophys J 778:17
111. Wise JH, Regan JA, O'Shea BW et al (2019) Nature 566:85
112. Woods TE et al (2019) Titans of the early universe: The Prato statement on the origin of the first supermassive black holes. Publ Astron Soc Aust 36(2019):e027
113. Wu X-B, Wang F, Fan X et al (2015) Nature 518:512
114. Wyithe JSB, Loeb A (2012) Mon Not R Astron Soc 425:2892
115. Zwick L, Mayer L, Haemmerlé L, Klessen RS (2022). arXiv e-prints arXiv:2209.02358

Chapter 2
Methods

Abstract In order to model GR instability supernovae, we first perform evolutionary simulations with a post Newtonian stellar evolution code HOSHI, both for metal free and metal enriched progenitors. This code includes a small nuclear network and neutrino cooling. We evolve these simulations to the general relativistic radial instability. We locate the instability by performing a normal mode analysis of the radial perturbations of the star in general relativity by linearizing the equation of motion. Next, when the star becomes unstable according to the general relativistic stability analysis, we simulate the dynamics of the explosion or collapse to a black hole by transporting the model to a general relativistic Lagrangian hydrodynamics code with a nuclear network. We use nuclear networks of different sizes depending on the progenitor, but the two main explosion mechanisms for these stars are the explosive alpha process and the CNO cycle and rapid proton capture process. Finally, if the model explodes, we compute the nucleosynthetic ejecta using post processing of the hydrodynamical trajectories. We also calculate the light-curve of the explosion using a radiation hydrodynamics code.

Keywords Stellar evolution · General relativistic radial instability · Hydrodynamics

In order to study SMSs and their subsequent explosions as GRSNe, we utilize several numerical and analytic techniques (Fig. 2.1). This approach is required because of the different timescales and physics involved during different parts of the star's lifetime and the explosion. Of course, a more consistent technique would utilize only one code for all parts, but this is numerically intractable with current computational resources.

The first of these techniques is a stellar evolution code HOSHI (Sect. 2.1). While we evolve the models in HOSHI, we check the stability of these models against the GR radial instability (Sect. 2.1.5). When the model becomes unstable due to the GR instability, we transport that profile from HOSHI to a GR hydrodynamics code. This code includes energy addition due to nuclear reactions and weak reactions, and is thus able to determine whether or not the unstable model will explode in a GRSN or collapse to a black hole (Sect. 2.2). Finally, for the exploding models, we transport them to an open source radiation hydrodynamics code (SuperNova Explosion Code, SNEC) where we calculate the light-curves of the GRSN (Sect. 2.3).

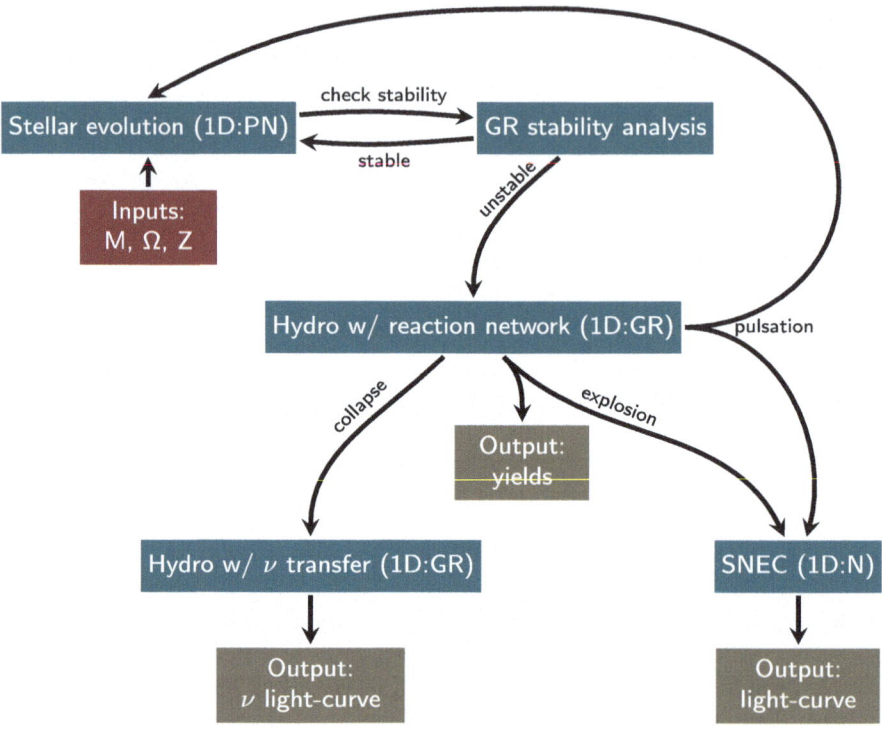

Fig. 2.1 Schematic diagram of the modeling of GRSNe

2.1 Stellar Evolution

HOSHI (HOngo Stellar Hydrodynamics Investigator) is a massive star evolution code, solving the stellar structure equations and at late times including the effects of hydrodynamics using a Henyey type implicit method [27–29, 34]. For stars larger than 1000 $M_\odot$, the small effects of general relativity play a large role in the stellar evolution, due to the stars being radiation dominated. [20] incorporated the first order static post Newtonian approximation to general relativity into the HOSHI code. We take thte convention that the isotopic mass excess, which is particularly large in the hydrogen envelope, is absorbed into the specific internal energy, thus bringing the post-Newtonian pressure gradient into very good agreement with the general relativistic pressure gradient. Despite this agreement, HOSHI's lack of a shock capture scheme and the dynamical corrections necessitate the use of another code. HOSHI's equation of state includes photons, averaged nuclei, electrons, and positrons. HOSHI uses the Rosseland mean opacity of the OPAL project [13] and solves the Saha equation to determine the ionization of hydrogen, helium, carbon, nitrogen, and oxygen. It also include neutrino cooling, mass loss, rotation, and a nuclear reaction network.

2.1 Stellar Evolution

In this thesis, M is the total mass, R the total radius, m_r the enclosed mass inside radius r, T the temperature, P the pressure, ϵ the internal energy, Γ_1 the local adiabatic index at constant entropy, and ρ_b the baryonic density where quantities with c subscripts show the central values. s is the entropy and s_r is the entropy due to radiation at a given mass [25]

$$s_r = 0.942 \left(\frac{M}{M_\odot} \right)^{1/2}. \tag{2.1}$$

Finally, X is the mass fraction of a specified element.

To assist with the analysis, we define various global energy quantities. The internal energy is

$$E_{\text{th}} = \int_0^M \epsilon \, dm_r, \tag{2.2}$$

where m_r is the mass coordinate and ϵ is the specific energy. The gravitational energy is

$$E_{\text{grav}} = -\int_0^M g_{\text{effective}} \, r \, dm_r, \tag{2.3}$$

where $g_{\text{effective}}$ is the local gravity with the 1st order PN correction to the static terms [22]. The accuracy of this approximation degrades with increasing density and velocity, neither of which are particularly concerning for our purposes. The kinetic energy is

$$E_{\text{kin}} = \int_0^M \frac{v^2}{2} \, dm_r, \tag{2.4}$$

where v is the radial velocity. The binding energy of the star is the negative of the thermal and gravitational energies (so that a more tightly bound star has higher E_{bind}), while the total energy additionally includes kinetic energy:

$$E_{\text{bind}} = -(E_{\text{th}} + E_{\text{grav}}) \tag{2.5}$$

$$E_{\text{tot}} = E_{\text{th}} + E_{\text{grav}} + E_{\text{kin}}. \tag{2.6}$$

As in our previous works, we define the explosion energy as the total energy at shock breakout. For the hydrodynamics code, we also report the integration over energy generation due to the nuclear network and neutrino cooling (dots indicate time derivatives):

$$E_{\text{nuc}}(t) = \int_0^t \int_0^M \dot{\epsilon}_{\text{nuc}} \, dm_r \, dt \tag{2.7}$$

$$E_\nu(t) = \int_0^t \int_0^M \dot{\epsilon}_\nu \, dm_r \, dt. \tag{2.8}$$

Next we discuss the differences between the metal free models presented in [22] and the metal rich ones presented in [19]. There are two salient points, related to initial conditions and to mass loss.

2.1.1 Initial Conditions

In both cases, we initiate the HOSHI code with a structure resembling an $n = 3$ polytrope and having a low central temperature ($T_c < 10^7$ K) and high entropy (that is, larger than the radiative value) as can be seen in Fig. 2.2. Entropies above the radiative value are expected for thermally supported supermassive protostars [11]. This low temperature, low density, polytrope then contracts towards the onset of nuclear burning.

In the case of metal free SMSs, this contraction continues well past the burning temperatures of normal massive stars. This is because the CNO cycle is required to support a SMS, due to its large mass. However, in a metal free environment, there is no carbon present to ignite the CNO cycle. Thus the star continues to contract until it reaches temperatures at which carbon can be synthesized via 3α reactions (Log $T_c \sim 8.2$). At this point, the SMS stabilizes and the central temperature can decrease by a small amount (to $T_c \sim 8.15$) during hydrogen burning. On the other hand, for metal rich SMSs, because of the presence of carbon, the CNO cycle can stabilize the star without requiring 3α reactions and the star instead stabilizes around Log $T_c = 7.7$. In either case, once nuclear burning begins, the entropy decreases to match the radiative value.

This procedure results, be design, in supermassive stars at the beginning of the main sequence. Due to their high entropies, however, supermassive stars are known to be able to ignite hydrogen burning while still accreting material [12]. This will thus result in a slight mismatch, particularly at the beginning of the stellar lifetimes, in the temperature profiles and chemical compositions of accreting and non accreting supermassive stars. Such differences can often be ignored, but for very large accretion rates, in particular the 10^4 $M_\odot$/yr inflows associated with the galaxy merger scenario [14, 15], they must be taken into account. Future work, both in investigating the accretion rates associated with these huge inflows and in modeling the stellar evolution of the resulting supermassive stars, is of interest. Accretion onto the star is generally thought to terminate with the exhaustion of the gas reservoir supplying the accreting material, though radiative feedback due to intermittent accretion may disrupt the flow Sakurai etal. 2015. Another scenario for producing accreting supermassive stars is runaway collisions in nuclear star clusters [6], though only the most extreme examples of these objects are likely to reach the GR instability.

2.1 Stellar Evolution

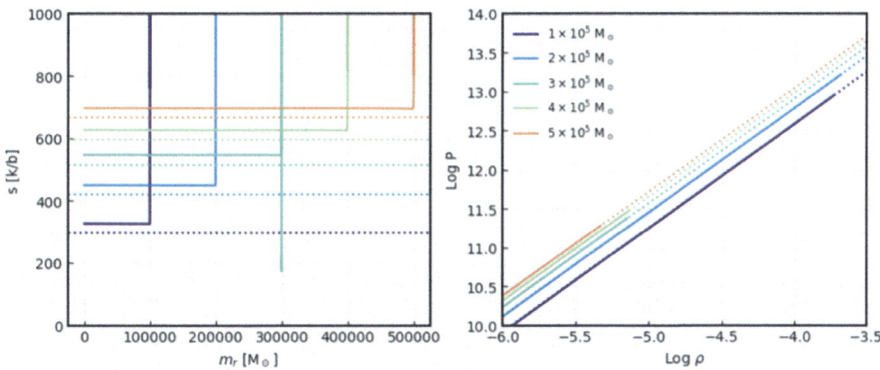

Fig. 2.2 Initial conditions in HOSHI for metal enriched SMSs of different masses. Left panel—entropy as a function of mass coordinate with radiative entropy shown by the dotted lines. Right panel—log pressure versus log density with $n = 3$ polytropic values shown by the dotted lines. Metal free SMSs have similar initial conditions at higher initial densities. Reprinted from [19], with permission from MNRAS

2.1.2 Mass Loss

When it comes to mass loss, the metal free and metal rich cases are treated identically in HOSHI, but fall in different regimes. SMSs are radiation dominated which suggests that mass loss may be important, especially for a rapidly rotating star [10]. However, in the case of metal free SMSs, the lack of metal lines with which a stellar wind could be driven means that mass loss is effectively zero (Fig. 3.3). On the other hand, for even modestly metal enriched SMSs, line driven mass loss can be very large and can effect the evolution and final fate of the SMS (Fig. 3.4). In this thesis, we adopt the prescription used in [34], but we will also show that for most masses and metallicities, mass loss does not overly effect the outcome. The exception to this is metal rich models on the lower end of the explosion mass range, which can lose enough mass to delay the onset of the GR instability.

2.1.3 Linearized Equation of Motion in GR

We now turn our attention to that very instability. In this subsection, we will derive the linearized equation of motion in spherical symmetry in GR following the derivation in [4]. We take the metric to be

$$ds^2 = -e^{2a}(cdt)^2 + e^{2b}dr^2 + r^2 d\Omega^2 \tag{2.9}$$

so that the 00, 11, and 01 field equations (where $\mu = 0, 1, 2, 3$ are the usual indices for t, r, θ, ϕ) can be written as

$$\frac{-8\pi G}{c^4} T_0^0 = \frac{-1}{r^2} \frac{\partial (re^{-2b})}{\partial r} + \frac{1}{r^2} \qquad (2.10)$$

$$\frac{-8\pi G}{c^4} T_1^1 = -e^{-2b} \left(\frac{2}{r} \frac{\partial a}{\partial r} + \frac{1}{r^2} \right) + \frac{1}{r^2} \qquad (2.11)$$

$$\frac{-8\pi G}{c^4} T_0^1 = \frac{-2e^{-2b}}{r} \frac{\partial b}{\partial t} \qquad (2.12)$$

where T_μ^ν is the Einstein tensor. First, we will derive the Tolman-Oppenheimer-Volkoff equation (TOV, [24] which describes a system in hydrostatic equilibrium, that is with no fluid motion. In this state, the diagonal terms of the Einstein tensor have a simple form

$$T_\mu^\nu (\mu = \nu) = (-\epsilon, P, P, P) \qquad (2.13)$$

such that the 00 field equation becomes

$$e^{-2b} = 1 - \frac{2G}{rc^2} \mathbb{M}_r \qquad (2.14)$$

where we have defined the general relativistic mass as

$$\mathbb{M}_r = \frac{4\pi}{c^2} \int_0^r \epsilon r^2 dr; \qquad (2.15)$$

note that this value does not match the baryonic mass coordinate m_r which is output by most simulations (there is a similar disagreement between density and baryonic density). Next, the 11 field equation becomes

$$\frac{e^{-2b}}{r} \frac{\partial 2a}{\partial r} = \frac{1}{r^2} (1 - e^{-2b}) + \frac{8\pi G P}{c^4}. \qquad (2.16)$$

We acquire one more equation by assuming the Einstein tensor is continuous ($\nabla_\mu T_\nu^\mu = 0$) and evaluate this condition in the radial direction,

$$\nabla_\mu T_1^\mu = \delta_\mu T_1^\mu + \Gamma^\mu_{1\mu\sigma} T_1^\sigma - \Gamma^\sigma_{1\mu 1} T_\sigma^\mu \qquad (2.17)$$

which becomes

$$\frac{\partial T_1^0}{\partial t} + \frac{\partial T_1^1}{\partial r} + T_1^0 \frac{\partial a + b}{\partial t} + (T_1^1 - T_0^0) \frac{\partial a}{\partial r} + \frac{2}{r} (T_1^1 - P) = 0 \qquad (2.18)$$

2.1 Stellar Evolution

after evaluation of the Christoffel symbols. This equation describes any configuration where the tensor is continuous, but for the case of hydrostatic equilibrium, we can set the off diagonal tensor terms to zero, and the expression reduces to:

$$\frac{\partial P}{\partial r} = -(P + \epsilon)\frac{\partial a}{\partial r} \qquad (2.19)$$

which can be substituted into the 11 field equation

$$\frac{-e^{-2b}}{2r(P+\epsilon)} \frac{\partial P}{\partial r} = \frac{1}{r^2}(1 - e^{-2b}) + \frac{8\pi G P}{c^4}. \qquad (2.20)$$

Finally, substitution of the 00 field equation (gravitational mass) yields the TOV equation

$$\left(1 - \frac{2G}{rc^2}\mathbb{M}_r\right)\frac{\partial P}{\partial r} = \frac{-1}{c^2}(P + \epsilon)\left(\frac{G}{r^2}\mathbb{M}_r + \frac{4\pi G P r}{c^2}\right). \qquad (2.21)$$

Imagine that we now perturb this configuration with a small Lagrangian displacement $v = \frac{\partial \xi}{\partial t}$ of the form $\xi(r) \propto e^{i\omega t}$, so that the primed quantities vary sinusoidally:

$$a \to a + a', \quad b \to b + b', \quad P \to P + P', \quad \epsilon \to \epsilon + \epsilon'. \qquad (2.22)$$

This assumption allows us to evaluate the full field equations if we only consider terms of first order, in order to keep the analysis tractable. Under this assumption, the off diagonal tensor components become

$$T_0^1 = -(P + \epsilon)v \qquad (2.23)$$

$$T_1^0 = e^{2b-2a}(P + \epsilon)v \qquad (2.24)$$

From here onwards, since we neglect terms of second order or greater, we emphasize that the results found below will always need to be verified with simulations. We now rewrite the field equations (Eqs. 2.10–2.12) and the continuity equation (Eq. 2.18) under the influence of this perturbation to first order

$$\frac{-8\pi G}{c^4}\epsilon' = \frac{-1}{r^2}\frac{\partial(re^{-2b}b')}{\partial r} \qquad (2.25)$$

$$\frac{-8\pi G}{c^4}P' = -e^{-2b}\left(\frac{2}{r}\frac{\partial a'}{\partial r} + \frac{b'}{r^2}\right) + \frac{b'}{r^2} \qquad (2.26)$$

$$\frac{-8\pi G}{c^4}(P + \epsilon)v = \frac{-2e^{-2b}}{r}\frac{\partial b'}{\partial t} \qquad (2.27)$$

$$e^{2b-2a}(P+\epsilon)\frac{\partial v}{\partial t} + \frac{\partial P'}{\partial r} + (P+\epsilon)\frac{\partial a'}{\partial r} + (P'+\epsilon')\frac{\partial a}{\partial r} = 0. \quad (2.28)$$

We will now attempt to eliminate the perturbed values other than ξ from these equations. First we will go after b'. We will integrate Eq. 2.27 over t to find

$$\frac{-8\pi G}{c^4}(P+\epsilon)\xi = \frac{-2e^{-2b}}{r}b' \quad (2.29)$$

Then, by subtracting the original 00 and 11 field equations from each other, we find

$$\frac{-8\pi G}{c^4}(P+\epsilon) = \frac{e^{-2b}}{r}\frac{\partial(2a+2b)}{\partial r} \quad (2.30)$$

so that we have

$$b' = -\xi\frac{\partial(a+b)}{\partial r} \quad (2.31)$$

This equation can be substituted into Eq. 2.25 to determine an expression for ϵ'

$$\epsilon' = \frac{-1}{r^2}\frac{\partial r^2\xi(P+\epsilon)}{\partial r} \quad (2.32)$$

$$\epsilon' = -\xi\frac{\partial \epsilon}{\partial r} - (P+\epsilon)\frac{e^a}{r^2}\frac{\partial(r^2 e^{-a}\xi)}{\partial r} \quad (2.33)$$

and we perform a similar substitution for Eq. 2.26 to determine an expression for a'

$$(P+\epsilon)\frac{\partial a'}{\partial r} = \frac{\partial(a+b)}{\partial r}\left(P' - (P+\epsilon)(2\frac{\partial a}{\partial r} + \frac{1}{r})\xi\right) \quad (2.34)$$

We can now plug expressions for the perturbed quantities into Eq. 2.28, one by one

$$e^{2b-2a}(P+\epsilon)\frac{\partial v}{\partial t} = -\frac{\partial P'}{\partial r} + \frac{\partial(a+b)}{\partial r}\left(P' - (P+\epsilon)(2\frac{\partial a}{\partial r} + \frac{1}{r})\xi\right) + (P'+\epsilon')\frac{\partial a}{\partial r} \quad (2.35)$$

$$e^{2b-2a}(P+\epsilon)\omega^2\xi = -\frac{\partial P'}{\partial r} + P'\frac{\partial(a+2b)}{\partial r} - \frac{\partial(a+b)}{\partial r}(P+\epsilon)(\frac{\partial 2a}{\partial r} + \frac{1}{r})\xi$$

$$+ \xi\frac{\partial \epsilon}{\partial r}\frac{\partial a}{\partial r} + (P+\epsilon)\frac{e^a}{r^2}\frac{\partial(r^2 e^{-a}\xi)}{\partial r}\frac{\partial a}{\partial r} \quad (2.36)$$

The final requisite equation is an expression for P'. We will not derive this equation here, but it follows from the conservation of baryon number [4].

2.1 Stellar Evolution

$$P' = -\xi \frac{\partial P}{\partial r} - \Gamma_1 P \frac{e^a}{r^2} \frac{\partial (r^2 e^{-a}\xi)}{\partial r} \quad (2.37)$$

After substitution of this equation and consolidation of terms (making use of the 22 field equation, for details see Sect. 6 of [4]), we arrive at the pulsation equation, where we have replaced ϵ with the relativistic density ρc^2, by convention:

$$-e^{-2a+2b}\omega^2 (P + \rho c^2)\xi = e^{-2a-b} \frac{d}{dr} \left[e^{3a+b} \Gamma_1 \frac{P}{r^2} \frac{d}{dr} (e^{-a} r^2 \xi) \right] - \frac{4}{r} \frac{dP}{dr} \xi$$

$$- \frac{8\pi G}{c^4} e^{2b} P(P + \rho c^2)\xi - \frac{1}{P + \rho c^2} \left(\frac{dP}{dr} \right)^2 \xi \quad (2.38)$$

Note that the Newtonian version of this equation (e.g. [25], Eq. 6.5.6) can be simply recovered by setting the second line to zero and substituting $a = b = 0$, $P \ll \rho c^2$.

2.1.4 Necessary Condition for Stability

A sufficient condition for instability is that there exists any ξ fulfilling the necessary boundary conditions which has an associated $\omega^2 < 0$. This indicates that instead of oscillation, the behavior of the star will be exponential in nature. This condition is useful if analytic solutions to the configuration exist, as is the case with a homogeneous or polytropic star. For a realistic numerical model, however, this condition is not particularly helpful. This is because testing several ξs and finding that the star is stable to those particular perturbations does not imply that the star is stable against all perturbations. One approach is to check one relatively simple perturbation which has a nearly linear dependence ($\xi \propto re^a$), and this approach already finds very different results from assuming a polytropic structure [9].

In this work, we construct a normal mode decomposition of the perturbations in the search for a specific perturbation, ξ_0. The sufficient condition for instability on ξ becomes a necessary condition on ξ_0. The crux of this approach is that Eq. 2.38 is self adjoint and thus solving it for ω can be treated as a Sturm-Liouville eigenvalue problem. Sturm-Liouville equations have the property that a sequence of solutions ξ_i exist such that the sequence forms a normal mode decomposition of all possible ξs; the i indicates the number of zero crossings in the solution. Each ξ_i is associated with an ω_i, and these frequencies posses the convenient property

$$\omega_0^2 < \omega_1^2 < \omega_2^2 < \cdots \quad (2.39)$$

Because of this property and the condition for instability, it is immediately clear that $\omega_0^2 < 0$ if and only if the star is unstable.

2.1.5 Evaluation of This Condition on Numerical Models

We adopt a straightforward numerical approach to solving the pulsation equation. For a given value of ω^2, we integrate Eq. 2.38 once from the center of the star to the surface to find one solution (ξ_{inner}) and then again from the surface to the center (ξ_{outer}). If $\xi_{\text{inner}} = \xi_{\text{outer}}$ continuously for some region of the star, then the combination of ξ_{inner} and ξ_{outer} constitute a valid to solution to Eq. 2.38.

We make an initial guess for the value of ω^2 and then compute ξ_{inner} and ξ_{outer}. In general, this will not be a solution, but we measure how close it is to being a solution by computing the Wronskian of ξ_{inner} and ξ_{outer} at a matching radius, p:

$$\mathcal{W}(p) = \xi_{\text{inner}}(p)\xi'_{\text{outer}}(p) - \xi_{\text{outer}}(p)\xi'_{\text{inner}}(p) \tag{2.40}$$

If the Wronskian is small, then the given ω^2 is near to a solution, but if it is large, then it is not.

We make use of this fact when iterating over the procedure in the previous paragraph, where we choose a subsequent guess for ω^2 based on an extrapolation of a function of the previous two guesses:

$$\mathcal{X}(p) = \frac{2\mathcal{W}(p)}{\xi_{\text{inner}}(p) + \xi_{\text{outer}}(p)}. \tag{2.41}$$

After some condition is reached (usually the Wronskian equaling zero within a certain number of decimals), we say that we have found a valid solution

$$\xi = \begin{cases} \xi_{\text{inner}}, & r \leq p \\ \xi_{\text{outer}}, & r > p \end{cases} \tag{2.42}$$

with frequency ω^2.

Next, we check to see which ξ_i we have found. In practice, we are interested only in ξ_0, but we can find any ξ_i with an appropriate initial guess for ω^2. Examples of the first 6 modes for an SMS and a poytrope are shown in Fig. 2.3. If we find a ξ_i which is not the fundamental mode, we then repeat the procedure with a lower initial guess for ω^2, but this is a rare case which usually arises from resolution problems in the stellar model.

Finally, we have recently [19] introduced a scheduled extrapolation (similar to a learning rate schedule in machine learning) which allows us to use smaller initial guesses which are dramatically more accurate.

$$\omega_0^2 = \frac{\omega_{0,1}^2 - \omega_{0,2}^2}{(\mathcal{X}_2 - \mathcal{X}_1)/\mathcal{X}_2} \times \begin{cases} 10^{-2} & \mathcal{X}_2/\mathcal{X}_{\text{initial}} > 10^{-4} \\ 10^{-1} & 10^{-4} > \mathcal{X}_2/\mathcal{X}_{\text{initial}} > 10^{-2} \\ 1 & \mathcal{X}_2/\mathcal{X}_{\text{initial}} < 10^{-2} \end{cases}. \tag{2.43}$$

2.1 Stellar Evolution

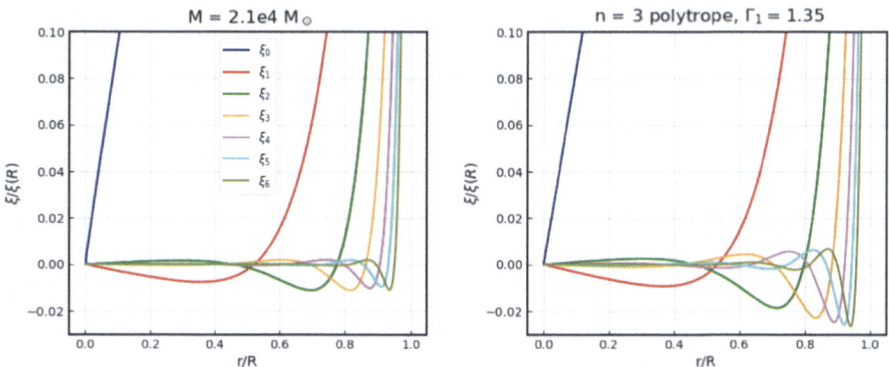

Fig. 2.3 First six modes of the perturbation for the first HOSHI timestep in the 2.1×10^4 M$_\odot$ metal free model (left panel) and an $n = 3$ polytrope (right panel). The perturbations have been normalized to $\xi_n(R) = 1$. Reprinted from [22], with permission from MNRAS

See Appendix A of [19] for details.

Before applying this stability analysis to SMSs, we first test it on numerical polytropes created by integrating the Lane-Emden equations using an RK4 integrator. The stability of polytropic models is well known to be dependent on Γ_1, specifically as [4]:

$$\Gamma_1 = \frac{4}{3} + \frac{2GMK}{Rc^2} \qquad (2.44)$$

for a numerical constant K dependent on polytropic index n. We put a sequence of $n = 3$ polytropes with Γ_1 from Eq. 2.44 and increasing numerical resolution into our stability analysis. We expect that it should return $\omega_0^2 = 0$, so the actual value which is returned can be regarded as an error associated with the analysis. Figure 2.4 shows that this error is small $\omega_0^2 \sim 10^{-13}$ compared to typical values for SMSs $\omega_0^2 \sim 10^{-6}$. This figure also shows that for increasing resolution, the error decreases dramatically. We can thus be confident that the stability analysis is working accurately, but we caution the reader that this is only a linear analysis and the final say regarding stability should rest with a GR hydrodynamics code.

Finally, we apply the stability analysis to HOSHI models of SMSs. When the star becomes unstable depends on many factors, but chief among them is the mass and metallicity. More massive models become unstable earlier in their lives, while more metal enriched models survive longer, due to their puffier structure. A typical case can be seen in Fig. 2.5 where $\omega_0^2 \sim 10^{-4}$ during hydrogen burning, before dropping down to $\omega_0^2 \sim 10^{-6}$ during helium burning, at which point the star becomes unstable. Note that typically, the model will react to the instability, and so appears to oscillate between stability and instability. However, this usually is a true instability which results in a collapse on much shorter timescales ($\sim 10^5$ s, compared with $dt_{\text{HOSHI}} \sim 10^9$ s) and we confirm this with the GR hydrodynamics code.

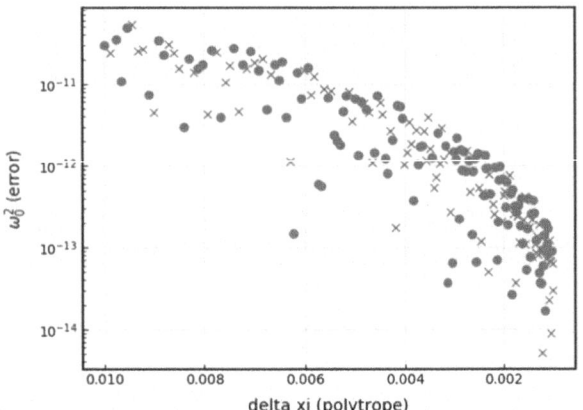

Fig. 2.4 Convergence of the accuracy of the GR stability criterion for numerical $n = 3$ polytropes as a function of polytrope spacing. Crosses are positive values of ω_0^2 and circles negative ones. Reprinted from [19], with permission from MNRAS

In some cases, the result of the hydrodynamics code is a pulsation rather than a collapse or an explosion. In this case, we port the model back to HOSHI and evolve the remnant. During this time, we again run the stability analysis. The star is usually stable during the quasi-static contraction, but then becomes unstable when approaching the equilibrium temperature (Fig. 2.6).

2.1.6 Accretion

In order to model the effects of accretion, we make the simple assumption that the accreted material has the same composition and entropy of the outermost 1% of the mass of the SMS. The former of these assumptions is valid because the central convective region never approaches the outer envelope, so the composition of the outer envelope will simply be the initial composition of the star (Pop III, solar metallicity, etc.). The latter assumption will ultimately depend on the details of the accretion, for instance if angular momentum is extracted from the flow by a disk or some other method. We view the constant entropy assumption as both straightforward and reasonably motivated. The evaluation of the GR instability in accreting SMSs is exactly the same as in non accreting SMSs. We note that the use of the stability analysis gives us lower final masses for the accreting SMSs than other studies which do not employ this analysis.

2.2 Hydrodynamics

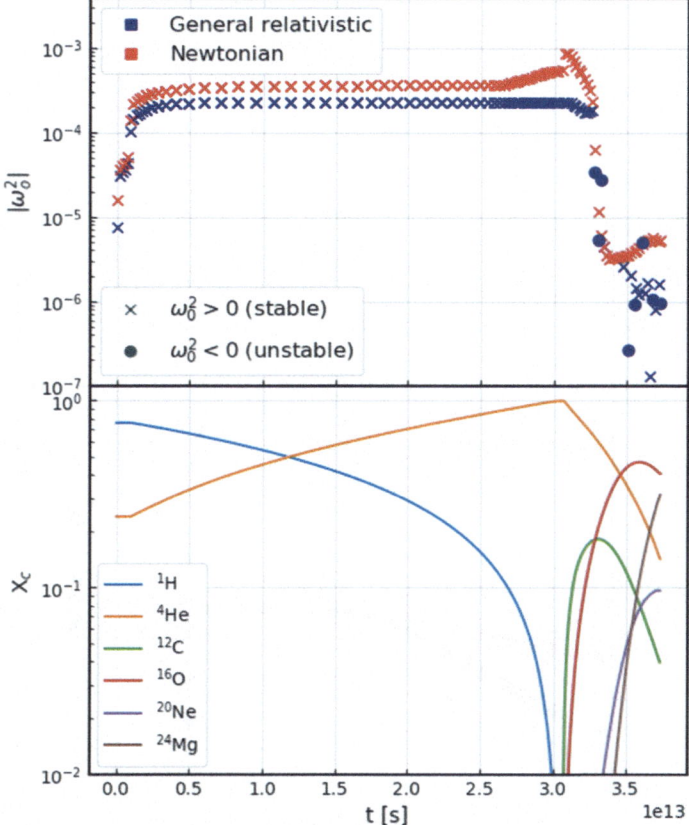

Fig. 2.5 GR stability analysis applied to the results of the 3×10^4 $M_\odot$ metal free HOSHI model. Upper panel—amplitude of the fundamental mode frequency as a function of time for the General Relativistic (blue) and Newtonian (red) analyses. Unstable models, having a negative value of frequency squared, are denoted by filled circles while stable models are denoted by crosses. Lower panel—Central mass fraction of various isotopes as a function of time. Reprinted from [22], with permission from MNRAS

2.2 Hydrodynamics

2.2.1 Numerical Scheme

The hydrodynamics code is a 1D GR implicit Lagrangian code [20, 21, 27–29, 34] based on the nuRADHYD code [21, 26, 33] which additionally solves neutrino transport. the hydrodynamics code uses a Roe-type approximate linearized Riemann solver. The independent variables are density, velocity, internal energy, entropy, electron fraction, radius, baryonic mass, enthalpy, and the two components of the Misner Sharp metric [16]. Internal energy changes primarily via energy generation from a

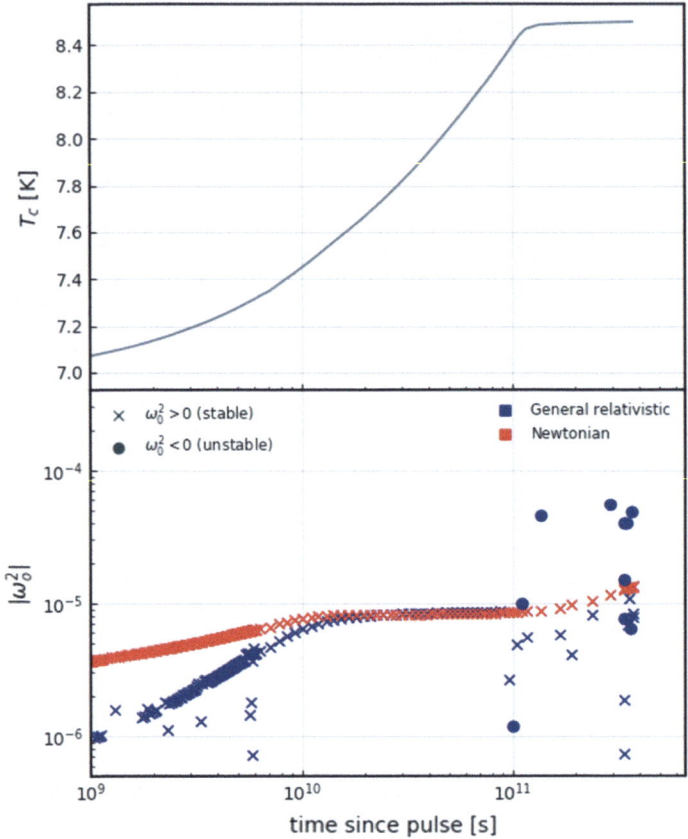

Fig. 2.6 GR stability analysis applied to the results of the 2.6×10^4 $M_\odot$ metal free model HOSHI model after the first pulse in the hydrodynamics code. Upper panel—evolution of central temperature in HOSHI. The Kelvin Helmholtz timescale is order 10^{11} s. Lower panel—GR stability analysis. The star becomes unstable near the end of the quasi-static contraction. Reprinted from Nagele et al. (2023b), with permission from MNRAS

nuclear network or cooling from thermal neutrino reactions. The equation of state is the same as in HOSHI.

In order to transport a model from HOSHI to the hydrodynamics code, we first determine the radial values of the the hydrodynamics code mesh according to the frequency function in Appendix B of [27]. This method is used to provide appropriate resolution to both the core and envelope of the SMS. Other variables are then inferred using linear radial interpolation of the log values of those variables, which we have found is the most accurate method of ensuring the fidelity (as a function of enclosed mass) of the remapping. Care must be taken at this step so as not to introduce unphysical perturbations. This is the same reason that we introduced the relativistic energy density to HOSHI (Sect. 2.2.3). However, an unavoidable coordi-

2.2 Hydrodynamics

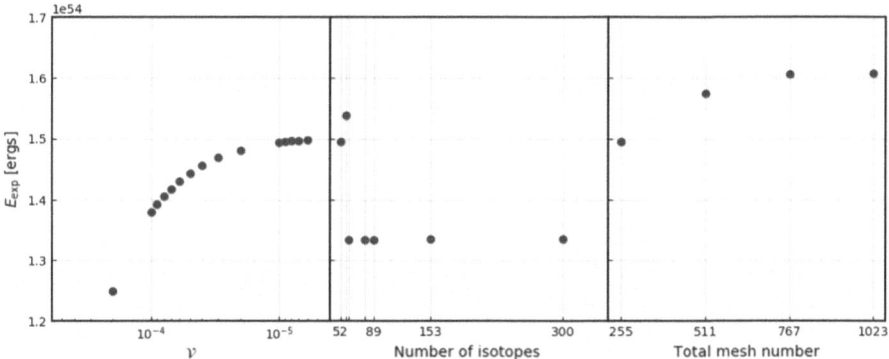

Fig. 2.7 Numerical convergence of explosion energy for three parameters of the hydrodynamics code for the 3×10^4 $M_\odot$ metal free model. Left panel—dependence on $\mathcal{V}$ (Eq. 2.45) with total mesh point number = 255 and isotope number = 52. Middle panel—dependence on number of isotopes in the nuclear network (52, 58, 61, 79, 89, 153, 300 with total mesh point number = 255 and $\mathcal{V} = 5 \times 10^{-4}$. Right panel—dependence of explosion energy on number of mesh points, with $\mathcal{V} = 5 \times 10^{-4}$ and isotope number = 52. Reprinted from [22], with permission from MNRAS

nate perturbation due to numerical error will always be present when changing mesh definitions, and this is a potential source of error in our simulation.

We continue the calculations for the explosions until there are convergence problems due to poor spatial resolution in the outer regions of the star. This typically occurs when the shock reaches 10^{15-16} cm. For the pulsations, a similar stopping condition is reached, but in this case we excise the ejected material from the simulation and verify that the remaining material becomes hydrostatic. For the models which collapse, we continue the calculation to a central temperature of 1 MeV, at which point neutrino heating begins to play a role.

In order to set the numerical parameters for the hydrodynamics code, we perform several numerical convergence tests using the the explosion energy, which is defined as the total energy when the shock reaches the stellar surface. The first panel of Fig. 2.7 shows the explosion energy as a function of $\mathcal{V}$, the limit on the maximum variation of the independent variables per times-step

$$\mathcal{V}(k) > \max_{i,j} \left| \frac{x_i(j, k-1)}{x_i(j, k)} \right|^{\pm 1} \quad (2.45)$$

where x_i is one of the independent variables of the hydrodynamics code, j is the mesh point number, and k is the time-step. The subsequent panels show the explosion energy as a function of total isotope number and mesh point number. If the limit in Eq. 2.45 is violated for a time-step k, that time-step is repeated with a reduced value of dt. Thus, a smaller $\mathcal{V}$ will require more time-steps which increases the time resolution of the simulation, making it more physically accurate. The simulations with greater resolution tend to reach slightly higher temperature, which correlates to an increase

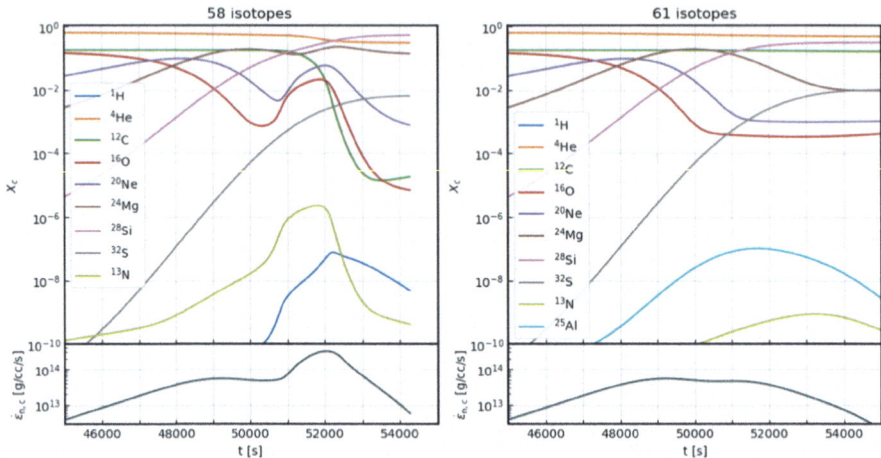

Fig. 2.8 Upper panel—comparison of central isotope mass fractions for the 3×10^4 $M_\odot$ metal free hydrodynamics model with a 58 isotope model (left) and a 61 isotope model (right). The high mass fraction of protons and ^{13}N in the left panel facilitates artificially high carbon burning. Lower panel—time evolution of central $\dot{\epsilon}$. Reprinted from [22], with permission from MNRAS

in E_{exp}. For the metal free models, we use $\mathcal{V} = 10^{-5}$, a 61 isotope network, and 767 mesh points. The 61 isotope network includes the p-p chain, CNO and hot CNO cycles, triple alpha, detailed alpha process until silicon, a more basic alpha process up to nickel, and photodissociation of heavy elements. For the metal rich models, we use 255 mesh points and $\mathcal{V} = 10^{-4}$ and a 153 isotope network which covers the low temperature rp process. This reduction in accuracy is due to the increased numerical cost of the expanded network, which is itself required by the more complex nature of the nucleosynthesis during the explosion.

2.2.2 Nuclear Networks

Throughout this thesis, we utilize a wide variety of nuclear networks. In this subsection, we give a brief overview of the motivation for these different networks and some of the challenges inherent in using smaller networks. We will first describe the metal free case and then the metal rich case.

In the metal free case, the explosion energy depends on $\mathcal{V}$ and mesh point number in straightforward ways, but the dependence on isotope number requires explanation. The main driver of the explosion is alpha capture reactions, but not all of these reactions proceed at the same rate. In particular, the carbon alpha capture rate is lower than other alpha capture reactions; note that we use 1.5 × the rate of [3], but have verified that the explosion energy depends very weakly on this reaction rate. $\mathcal{O}(10\%)$ of the mass of the star is carbon, and very little of this is burned via carbon

2.2 Hydrodynamics

alpha capture on the timescale of the explosion. However, if nucleons are present, catalysis enhances the carbon alpha capture rate with

$$^{12}C(p, \Gamma_1)^{13}N, \ ^{13}N(\alpha, p)^{16}O. \tag{2.46}$$

In the networks with isotope number less than 61 (Fig. 2.8, left panel), a reservoir of free nucleons is built up during the explosion, and these then serve as the catalyst for carbon burning. In the networks with higher isotope numbers, however, the nucleons are absorbed in reactions such as

$$^{24}Mg(p, \Gamma_1)^{25}Al. \tag{2.47}$$

Indeed, aluminium is of particular importance (Fig. 2.8, right panel) because the inclusion of its isotopes is the only difference between the 58 isotope network and the 61 isotope network, and from Fig. 2.7, we can see that this is the isotope number where the explosion energy converges. Appendix B of [22] contains a steady state calculation verifying this explanation.

For the metal rich case, the main drivers of the explosion are the CNO cycle (and other cyclical reactions) as well as the rp process. The rp process, in particular, is difficult to implement because of the large number of isotopes which could become involved. We take a two step approach to this problem, coupling the hydrodynamics to a 153 isotope network which covers the p side up to zinc at varying depths. This network is able to capture the reactions which contribute to the vast majority of energy production during the explosion, but it does suffer from edge effects, thus requiring post processing for accurate nucleosynthetic yields. After the simulation is complete, we post process with a 514 isotope network. This includes elements up to ruthenium, with coverage over the p side inside the line with slope unity going through ^{14}F (inclusive). A summary of all networks can be found in Table 2.1.

2.2.3 Consistency with Evolutionary Profiles and Stability Analysis

After first implementing the stability analysis, we found that models which were deemed to be stable would still collapse in the hydrodynamics code. Upon further investigation, we discovered a deviation in the pressure gradient between the evolutionary models and the hydrodynamical models. Although the evolutionary models included the effects of general relativity via the 1st order post Newtonian approximation to the TOV equation (Eq. 2.21), this effect was being incorrectly implemented, and we discovered that this had also been the case in previous papers [5, 20]. The first order approximation should result in a pressure gradient which is consistent to one part in 10^5, with the second order approximation consistent to one part in 10^7 and so on. However, we found that the numerical models were only consistent to

Table 2.1 Summary table for nuclear networks, indicated at the top by the total number of isotopes. Entries show the range in A for the specified element

Element	52	58	61	153	300	514
n	1	1	1	1	1	1
p	1–3	1–3	1–3	1–3	1–3	1–3
He	3–4	3–4	3–4	3–4	3–4	2–4
Li	6–7	6–7	6–7	6–7	6–7	4–7
Be	7–9	7–9	7–9	7–9	7–9	5–9
B	8–11	8–11	8–11	8–11	8–11	6–11
C	12–13	12–13	12–13	12–13	11–16	8–14
N	13–15	13–15	13–15	13–15	13–18	10–16
O	14–18	14–18	14–18	14–18	14–20	12–18
F	17–19	17–19	17–19	17–19	17–22	14–20
Ne	18–20	18–20	18–20	18–22	18–24	16–22
Na	23	23	23	21–23	21–26	18–24
Mg	24	24–26	24–26	22–26	22–28	20–26
Al	27	27	25–27	25–27	25–30	22–28
Si	28	28–30	28–30	26–32	26–32	24–30
P	31	31	31	29–33	27–34	26–32
S	32	32–34	32–34	30–36	30–37	28–37
Cl	35	35	35	33–37	32–38	30–39
Ar	36	36	36	34–40	34–43	32–43
K	39	39	39	37–41	36–45	34–45
Ca	40	40	40	38–43	38–48	36–48
Sc	43	43	43	41–45	40–49	38–49
Ti	44	44	44	43–48	42–51	40–51
V	47	47	47	45–51	44–53	42–53
Cr	48	48	48	47–54	46–55	44–55
Mn	51	51	51	49–55	48–57	46–57
Fe	52–56	52–56	52–56	51–58	50–61	48–60
Co	55–56	55–56	55–56	53–59	51–62	50–61
Ni	56	56	56	55–62	54–66	52–66
Cu	–	–	–	57–63	56–68	54–68
Zn	–	–	–	60–64	59–71	56–71
Ga	–	–	–	–	61–73	58–73
Ge	–	–	–	–	63–75	60–75
As	–	–	–	–	65–76	62–76
Se	–	–	–	–	67–78	64–81
Br	–	–	–	–	69–79	66–82
Kr	–	–	–	–	–	68–86
Rb	–	–	–	–	–	70–87
Sr	–	–	–	–	–	72–89
Y	–	–	–	–	–	74–91
Zr	–	–	–	–	–	76–95
Nb	–	–	–	–	–	78–96
Mo	–	–	–	–	–	80–98
Tc	–	–	–	–	–	82–98
Ru	–	–	–	–	–	84–99

2.3 Supernova Lightcurves

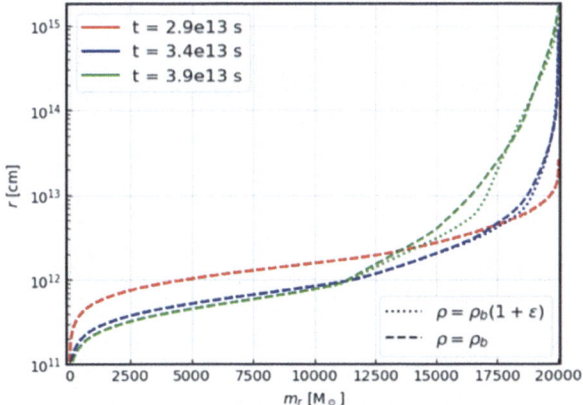

Fig. 2.9 Comparison of radial time snapshots for the 2×10^4 $M_\odot$ model in the HOSHI code with (dotted lines) and without (dashed lines) isotopic mass excess contributions to the internal energy. The inclusion of internal energy to the GR density when evaluating the PN TOV equation causes a more compact inner envelope, while the outer envelope remains largely unchanged. Reprinted from [22], with permission from MNRAS

one part in 10^2. This is because we had been using the baryonic density in Eq. 2.21, whereas we should have been using the relativistic density. This difference is particularly stark in the hydrogen envelope, where the isotopic mass excess is high. After rerunning the stellar evolution calculations using the relativistic density, we found more compact envelopes which were consistent with the pressure gradients in the hydrodynamics code (Fig. 2.9). These models were also more consistent with the results of the stability analysis.

2.3 Supernova Lightcurves

2.3.1 SNEC

SNEC [18] is an open source 1D Lagrangian hydrodynamics code which computes photosphere position and luminosity. From these quantities, we calculate effective temperature and assume a blackbody spectrum to determine AB magnitudes for JWST and other observatories. Our condition for porting to SNEC is that the explosion energy determined by SNEC must match that of the hydrodynamics code to 1% accuracy, because we expect the explosion energy from the hydrodynamics code to be accurate. In practice, this occurs once the shock has crossed into the region with small velocity ($\sim 10^{13}$ cm, Fig. 3.7) which is roughly determined by the sound speed multiplied by the explosion timescale. We follow the evolution of the explosion in SNEC until well past the end of the plateau phase identified in [17].

2.4 Supernova Rate Estimation

In this section, we will give estimates for the rate of GRSNe, both for metal free SMSs [31] and for the rapidly assembled metal rich SMSs in the galaxy merger scenario [15].

2.4.1 Metal Free

For metal free SMSs, the scenario which has been investigated in the most detail is the radiation induced formation scenario, and we base our estimates on the results of those investigations. The number density of DCBHs which result in SMSs with mass greater than 10000 $M_\odot$ has been estimated to lie anywhere in the range 10^{-1}–10^{-9} cMpc^{-3} [32]. Taking the very optimistic 10^{-2} cMpc^{-3} [1, 8] results in tens of billions of SMSs between redshifts 7 and 30, which spans a time period of just over half a billion years. The empirically derived mass dependence of SMSs in [30], i.e. $M^{-2.8}$ (Fig. 11 of [30]) suggests that only 1.4% of SMSs fall within the mass range of the explosions and pulsations investigated in this thesis, so that 500 million GRSNe will have went off in this redshift range and volume. This works out to a rate of roughly one GRSN per year in the rest frame, or about one per decade in the observer frame.

2.4.2 Metal Rich

[2] calculated the number density of massive galaxy mergers ($M_{\text{halo}} > 10^{11}$ $M_\odot$) which fulfilled the criteria for merger induced direct collapse (Fig. 4), which is order of magnitude $\phi = 10^{-4}$ [cMpc^{-3} Gyr^{-1}] (note that Fig. 4 of [2] has units of [cMpc^{-3} 0.1 Gyr^{-1}]). As discussed in [2], relaxing the mass asymmetry condition further would increase the merger rate by an order of magnitude. Another increase could be had by relaxing the mass constraint. Indeed, the supermassive stars considered in this thesis may be the result of mergers between slightly less massive galaxies, which are thought to occur more often (e.g. [23]). However, it should be noted that the rate assumed in [2] would severely overproduce SMBHs and can thus be regarded as an extreme upper limit on the metal rich GRSN rate. In a similar vein, a lower limit for this rate can be found by assuming one metal rich GRSN per high redshift quasar, which translates to a rate of 10^{-9} [cMpc^{-3} Gyr^{-1}] [e.g. 7].

References

1. Agarwal B, Khochfar S, Johnson JL et al (2012) Mon Not R Astron Soc 425:2854
2. Bonoli S, Mayer L, Callegari S (2014) Mon Not R Astron Soc 437:1576

References

3. Caughlan GR, Fowler WA (1988) Atomic Data Nucl Data Tables 40:283
4. Chandrasekhar S (1964) Astrophys J 140:417
5. Chen K-J, Heger A, Woosley S et al (2014) Astrophys J 790:162
6. Denissenkov PA, Hartwick FDA (2014) Mon Not R Astron Soc 437:L21
7. Fan X, Banados E, Simcoe RA (2022). arXiv e-prints arXiv:2212.06907
8. Habouzit M, Volonteri M, Latif M, Dubois Y, Peirani S (2016) Mon Not R Astron Soc 463:529
9. Haemmerlé L (2021) Astron Astrophys 647:A83
10. Haemmerlé L, Meynet G, Mayer L et al (2019) Astron Astrophys 632:L2
11. Hosokawa T, Omukai K, Yorke HW (2012) Astrophys J 756:93
12. Hosokawa T, Yorke HW, Inayoshi K, Omukai K, Yoshida N (2013) Astrophys J 778:178
13. Iglesias CA, Rogers FJ (1996) Astrophys J 464:943
14. Mayer L, Fiacconi D, Bonoli S et al (2015) Astrophys J 810:51
15. Mayer L, Kazantzidis S, Escala A, Callegari S (2010) Nature 466:1082
16. Misner CW, Sharp DH (1964) Phys Rev 136:571
17. Moriya TJ, Chen K-J, Nakajima K, Tominaga N, Blinnikov SI (2021) Mon Not R Astron Soc 503:1206
18. Morozova V, Ott CD, Piro AL (2015) SNEC: SuperNova Explosion Code. Astrophys Source Code Libr. record ascl:1505.033, ascl:1505.033
19. Nagele C, Umeda H, Takahashi K (2023). arXiv e-prints arXiv:2301.01941
20. Nagele C, Umeda H, Takahashi K, Yoshida T, Sumiyoshi K (2020) Mon Not R Astron Soc 496:1224
21. Nagele C, Umeda H, Takahashi K, Yoshida T, Sumiyoshi K (2021) Mon Not R Astron Soc 508:828
22. Nagele C, Umeda H, Takahashi K, Yoshida T, Sumiyoshi K (2022) Mon Not R Astron Soc 517:1584
23. O'Leary JA, Moster BP, Naab T, Somerville RS (2021) Mon Not R Astron Soc 501:3215
24. Oppenheimer JR, Volkoff GM (1939) Phys Rev 55:374
25. Shapiro SL, Teukolsky SA (1983) Black holes, white dwarfs, and neutron stars : the physics of compact objects
26. Sumiyoshi K, Yamada S, Suzuki H et al (2005) Astrophys J 629:922
27. Takahashi K, Sumiyoshi K, Yamada S, Umeda H, Yoshida T (2019) Astrophys J 871:153
28. Takahashi K, Yoshida T, Umeda H (2018) Astrophys J 857:111
29. Takahashi K, Yoshida T, Umeda H, Sumiyoshi K, Yamada S (2016) Mon Not R Astron Soc 456:1320
30. Toyouchi D, Inayoshi K, Li W, Haiman Z, Kuiper R (2022). arXiv e-prints arXiv:2206.14459
31. Valiante R, Agarwal B, Habouzit M, Pezzulli E (2017) On the formation of the first quasars. Publ Astron Soc Aust 34:e031
32. Woods TE, Agarwal B, Bromm V et al (2019) Titans of the early Universe: The Prato statement on the origin of the first supermassive black holes. Publ Astron Soc Aust 36:e027
33. Yamada S (1997) Astrophys J 475:720
34. Yoshida T, Takiwaki T, Kotake K et al (2019) Astrophys J 881:16

Chapter 3
Results

Abstract For both types of supermassive stars, metal free stars which explode via the explosive alpha process, and metal enriched stars, which explode via the CNO cycle and rapid proton capture process, we find the existence of large explosion windows and conclude that general relativistic instability supernovae are a general consequence of the existence of supermassive stars. These results run contrary to the results of previous papers which found that these types of explosions were much rarer. We also identify the existence of thermonuclear pulsations, which were previously unknown. Both of these advances are primarily due to the use of the general relativistic stability analysis. Finally, we compare our results to existing James Webb Space Telescope data, particularly NIRSpec observations of GN-z11.

Keywords Supermassive stars · Supernova lightcurves · Nucleosynthetic yields

This chapter has a similar layout to the last one, starting off with the results of the stellar evolution simulations, including the stability of the SMSs, and then proceeding to the outcomes of the hydrodynamical simulations and observable quantities.

3.1 Stellar Evolution

3.1.1 GR Stability

We first turn to the question of stability, and a comparison of our criterion to others'. First, and most common in the literature [8, 28, 32, e.g.] is the polytropic criterion, that is treating the SMS as having a polytropic structure. We will also compare our results to those of Haemmerlé [9].

SMSs have high entropy and are supported mostly by radiation pressure and this invites analytic approximations. In particular, it is often assumed that the SMS core is very nearly an $n = 3$ polytrope. The explosive α process explosion explored in Nagele et al. [21] involves helium burning, so we will calculate the instability condition for a polytrope consisting of pure helium, as a function of the mass of the helium core (M_{He}) which is taken from the stellar evolution calculation. Once the mass of the helium core is known, the radiation entropy may be approximated as $s_r \propto M_{He}^{1/2}$ [25]. From here, the polytropic constant is determined

$$K = \frac{a}{3}\left(\frac{3s_r}{4m_p a}\right)^{4/3}. \tag{3.1}$$

Next, we calculate the outer radius at which the star will be unstable, R_{crit} by setting the SMS Γ equal to the general relativistic Γ_1,

$$\frac{4}{3} + \frac{\beta}{6} + \mathcal{O}(\beta^2) = \frac{4}{3} + \frac{2GM_{He}\kappa}{R_{crit}c^2} \tag{3.2}$$

where $\beta \approx 4.3/\mu(M/M_\odot)^{-1/2}$ is the ratio of the gas pressure to total pressure, and $\kappa = 2.249$ for $n = 3$ (this form of κ differs from Chandrasekhar [4] by a factor of 2). Equation 3.2 can be solved for R_{crit} as a function of M_{He}, so that we finally arrive at an expression for the critical density [25]

$$\rho_{crit} = \left[\frac{R_{crit}}{\xi_1}\left(\frac{K}{\pi G}\right)^{-1/2}\right]^{-3} \propto M_{He}^{-7/2} \tag{3.3}$$

as a function of M_{He}, where $\xi_1 = 6.897$ is determined numerically. Thus, we have an expression for the critical density of a purely helium SMS core as a function of M_{He}. By construction, the SMS models in Nagele et al. [21] are near to this point according to the GR stability analysis. Figure 3.1 shows that, for these models, the central density when the star is pure helium is more than an order of magnitude below the critical density. This shows that the polytropic criterion underestimates the GR instability in comparison to our analysis.

We now compare our method to the results of Haemmerlé [9], who also evaluate Eq. 2.38, though they make the assumption $\xi \propto re^a$ instead of finding ξ_0. Figure 3.2 shows the helium mass fraction at the first instability—that is, the first model in the HOSHI calculation which is unstable—determined by either method. The central helium mass faction for an explosion or pulsation is roughly $0.1 < X_c(^4He) < 0.7$, c.f. Table 3.3. Our method uses a necessary condition for instability while the method of Haemmerlé [9] uses a sufficient condition. This means that there are models which would be stable according to the method of Haemmerlé [9], but not according to our method. Furthermore, any model which is unstable according to the method of Haemmerlé [9] will also be unstable according to our method. Thus, our method will find an instability earlier in the evolutionary calculation than the method of Haemmerlé [9]. The longer the time before the instability in the evolutionary calculation, the higher the mass range for the explosion, because lower mass models which would explode if the instability occurred earlier instead burn all of their helium ($M < 3 \times 10^4$ in Fig. 3.2). This means that the explosion would occur at larger mass, if we were to use the method of Haemmerlé [9], although that being said, the explosion mass range found by their method would still be significantly smaller

3.1 Stellar Evolution

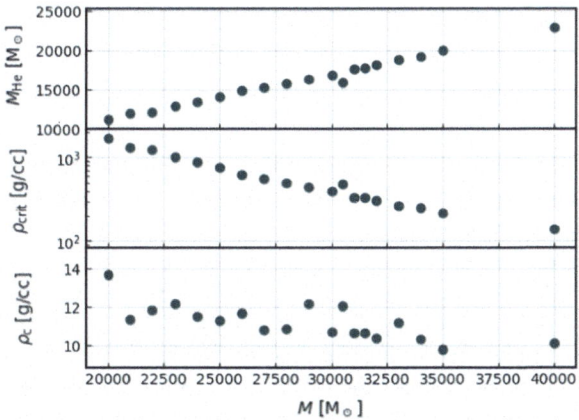

Fig. 3.1 Evaluation of the polytropic criterion for for the supermassive stars in Nagele et al. [21]. Upper panel—helium core mass at the GR instability as determined by the stability analysis. We take the envelope to be defined as the mass outside of $X(^1H) > 1e-5$. Middle panel—evaluating the polytropic criterion on the helium cores from the upper panel. Lower panel—central densities of the HOSHI models. A comparison to the middle panel yields the conclusion that these models are not yet unstable according to this criterion, even though we know that they are unstable (from the stability analysis and hydrodynamics). Reprinted from Nagele et al. [21], with permission from MNRAS

Fig. 3.2 Comparison of the first instability reached by two methods, Sect. 2.1.5 and that of Haemmerlé [9]. Reprinted from Nagele et al. [21], with permission from MNRAS

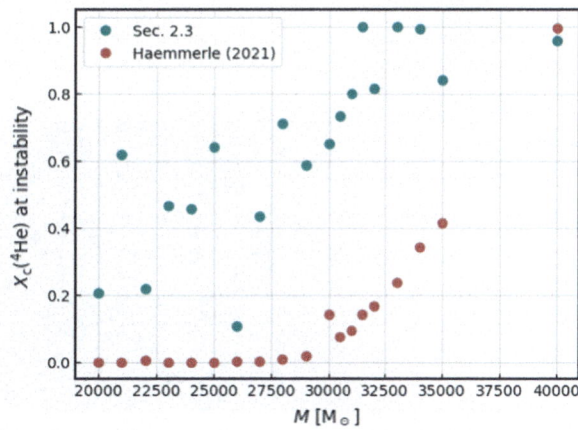

than the case of $M \sim 5 \times 10^4$ $M_\odot$ discussed in Chen et al. [6, 20]. In summary, our method of evaluating the GR radial instability finds an earlier instability than both the polytropic criterion and Haemmerlé [9], which tends to shift the mass range for the explosion downwards.

3.1.2 Metal Free

Unlike normal stars, SMSs cannot sustain hydrogen burning with just the p-p chain [31], and they instead require the CNO cycle. The lack of metals in primordial SMSs thus results in a peculiar scenario whereby the central temperature of the star must first rise to around Log $T_c = 8.2$ so that carbon can be synthesize via the triple alpha reaction. Once even a small amount (typically $X \sim 10^9$) is synthesized, the CNO cycle turns on and the star enters the hydrogen burning phase. At this time, the star is completely convective (Fig. 3.3) though as hydrogen burning progresses, radiative regions form near the surface and a helium core is eventually formed. If the SMS survives to helium burning but does not survive until oxygen burning, it is this core that provides the fuel for the α process explosion. The envelope forms its own convective structure which is stochastic with regards to the parameters of the simulation. This envelope structure can affect the stability of the SMS or even its explodability [20], thus necessitating our stability analysis.

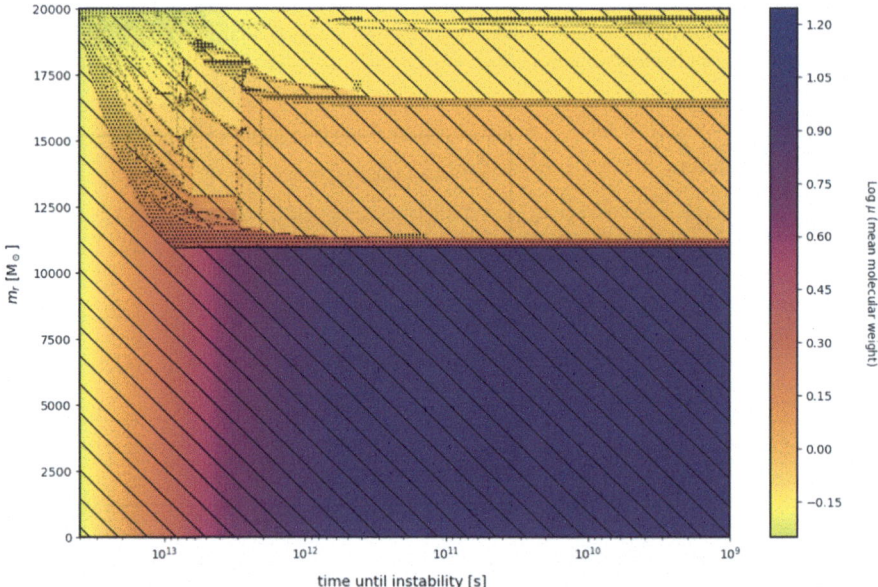

Fig. 3.3 Kippenhahn diagram for the 2×10^4 $M_\odot$ metal free model, with time until the instability on the horizontal axes, and mass coordinate on the vertical axis. The color shows the log of the mean molecular weight. Diagonal hatches show convective regions, dotted hatches show non convective regions, and cross hatches show semi-convective regions. Note that the time-step of HOSHI at this point in the evolution is roughly 10^9 s. Reprinted from Nagele et al. [21], with permission from MNRAS

3.1 Stellar Evolution 41

3.1.3 Metal Rich

In this subsection, we will first discuss four specific models with $Z = 0.1\ Z_\odot$ which are relevant to Sect. 3.5, and can be treated as illustrative examples. We will then turn to a more general discussion of metal rich SMSs.

Unlike in the metal free case, the metal rich models, even those with sub-solar metallicity, contain the material required for the CNO cycle. Thus, although these stars experience high temperatures at the beginning of hydrogen burning, they do not reach the same extremes as in the previous subsection. Similarly to the metal free models, however, the metal rich models are completely convective at the start of hydrogen burning (Fig. 3.4). Radiative surface layers then develop during late hydrogen burning, ultimately leading to core formation, though the most massive model in the four examples becomes GR unstable before this. Just before hydrogen depletion, strong shell burning begins in the radiative region, soon after which the M= $5 \times 10^4\ M_\odot$ becomes GR unstable, with a central hydrogen mass fraction of $\sim 10^{-6}$. The two low mass models survive the transition to helium burning and eventually form carbon-oxygen cores. During this stage, the hydrogen shell burning intensifies, resulting in a largely convective envelope. In addition, helium shell burning begins in a thin layer just below the hydrogen shell burning region. During this period, dredge up of helium, carbon, and oxygen from the core is efficient, eventually resulting in super-solar abundances of carbon and oxygen, which we will revisit in Sect. 3.5.

In general, the fate of metal enriched models depends on the initial mass and metallicity. We categorize these models by the evolutionary stage during which they become unstable: after hydrogen burning, during hydrogen burning, and before hydrogen burning. Models that collapse after hydrogen burning are either low mass or are mass loss models at solar metallicity. These solar metallicity mass loss models lose large amounts of mass even during hydrogen burning, which delays the onset of the GR instability, simply due to the lower total mass of the star. In many cases, the mass loss means that the star does not become GR unstable before the end of the HOSHI calculation. The one metal enriched model that becomes unstable in between hydrogen and helium burning collapses due to lack of fuel ($5 \times 10^4\ M_\odot$, $0.1\ Z_\odot$) and we did not find any models which become unstable during helium burning.

At sub-solar metallicities, mass loss is weaker and most of the models become unstable during hydrogen burning. However, the exact moment of the onset of the GR radial instability does not follow any obvious trends in mass or metallicity. Such trends may exist, but would require much larger numbers of stellar models. In general, the stellar structure is a function of the complex interplay between energy generation, opacity, and convection, and each of these depends on the metallicity of the star in nontrivial ways. Furthermore, convection is known to depend on the initial mass of the star in a stochastic manner. This complexity does not change the fact that all models which collapse during the hydrogen burning phase will explode if they have sufficient metallicity (Sect. 3.2.2).

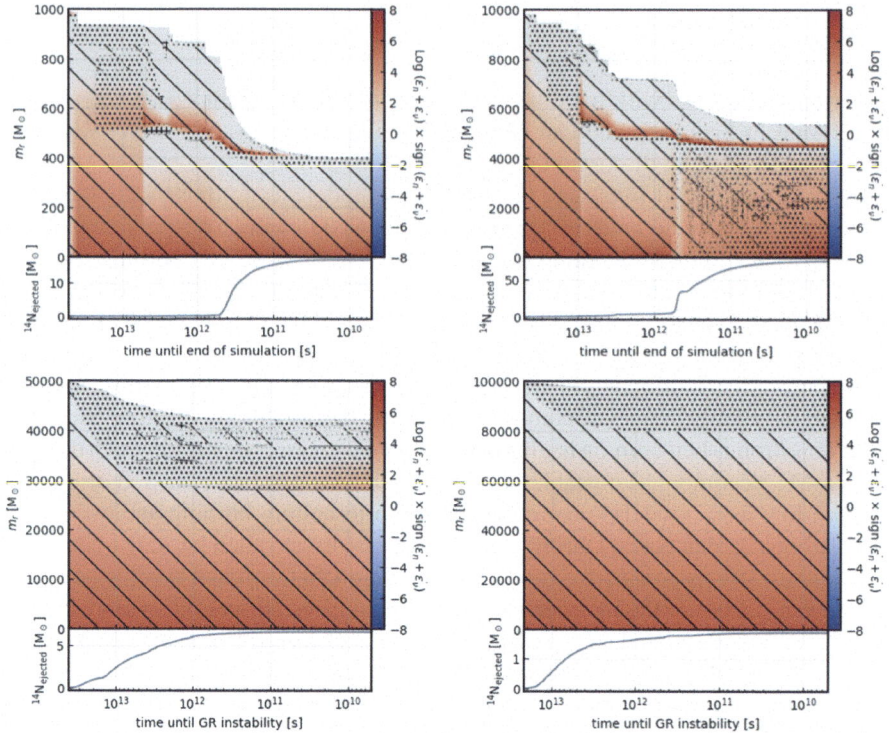

Fig. 3.4 Kippenhahn diagrams for four metal enriched HOSHI models. The horizontal axis is time until the end of the simulation (for the two lower mass models) or until the GR instability (for the two higher mass models) while the vertical axes are mass coordinate and cumulative ^{14}N ejected from the stellar wind. In the upper half of each panel the color shows the heating and [very small] cooling rate from nuclear and neutrino reactions. The hatches show convective (diagonal), radiative (dotted) and semi-convective (crossed) regions. Reprinted from Nagele et al. [16] under the Creative Commons Attribution 4.0 licence

Finally we have models which are GR unstable from the very beginning of the simulation. These models are discussed in Nagele et al. [18], but are less interesting now that the code has been modified to include accretion (Table 3.1).

3.1.4 Accreting

The accreting models begin the stellar evolution as 1000 $M_\odot$ pre-main sequence seeds. After a period of rapid contraction, they settle into an equilibrium state while they accrete more material. Figure 3.5 shows four Kippenhahn diagrams for four of the Pop III models with different accretion rates. A striking feature which can be seen immediately is that the ratio of the convective core to the total mass decreases for

3.1 Stellar Evolution

Table 3.1 Quantities related to the stellar models at the GR instability. The columns are, initial mass, metallicity, and mass loss, followed by several quantities at the GR instability: mass, radius, central temperature, central density, central entropy and its ratio to the radiative value, central hydrogen mass fraction, and binding energy

M [10^5 $M_\odot$]	$Z/Z_\odot$	$\dot{M}$	M_f [10^5 $M_\odot$]	R [10^{14} cm]	T_c [10^7 K]	ρ_c [10^{-3} g/cm^3]	s_c [kb/b]	s_c/s_r	X(^{1}H)	E_{bind} [10^{54} ergs]
0.5	1	yes	0.3468	1.387	6.577	202.2	186.4	1.063	0.1262	0.2759
0.5	10^{-1}	yes	0.4223	1.602	13.76	1789	188.5	0.9735	2.277e$^-$6	0.5161
0.5	10^{-2}	yes	0.4923	10.85	12.21	1220	192.9	0.9232	0.001927	0.4976
1.0	1	no	1	1.453	6.502	119.2	297.7	0.9993	0.2005	0.7939
1.0	10^{-1}	no	1	1.012	7.387	173.6	301.8	1.013	0.3003	1.032
1.0	10^{-1}	yes	0.9718	2.856	7.186	158.7	306.8	1.045	0.4186	1.154
1.0	10^{-2}	yes	0.9963	0.8019	8.481	259.4	307.7	1.035	0.4009	1.362
1.0	10^{-3}	no	1	0.6387	9.925	411.7	311.2	1.045	0.4441	1.693
1.0	10^{-3}	yes	0.9997	0.5608	9.812	395.4	314.6	1.056	0.5148	1.796
1.0	2×10^{-2}	yes	0.9923	2.049	8.212	238.9	301.9	1.017	0.3222	1.185
1.0	4×10^{-2}	yes	0.9841	2.075	7.744	199.2	304.7	1.031	0.3688	1.165
1.0	6×10^{-2}	yes	0.9529	2.628	7.737	204.2	294.3	1.012	0.2566	0.986
1.0	8×10^{-2}	yes	0.9408	3.017	7.481	184.1	295.6	1.023	0.2764	0.9706
1.1	10^{-1}	yes	1.036	2.598	7.398	169	311	1.025	0.3012	1.071
1.2	10^{-1}	yes	1.169	1.276	7.236	147.3	335.6	1.042	0.4376	1.412
1.3	10^{-1}	yes	1.277	0.8369	7.153	135.1	354.8	1.054	0.5361	1.699
1.5	1	no	1.5	1.101	6.279	84.66	378.4	1.037	0.4213	1.538
1.5	1	yes	1.077	0.5428	6.565	112.8	322.3	1.042	0.2021	0.908
1.5	10^{-1}	yes	1.467	0.9309	7.226	130.1	377.2	1.045	0.498	1.877
2.0	1	no	2	3.016	1.917	2.06	448.6	1.065	0.7154	0.8276
2.0	10^{-1}	yes	2	0.9452	2.97	7.659	448.6	1.065	0.7555	1.33
2.5	1	no	2.5	4.451	1.352	0.6466	499	1.059	0.7154	0.7215
2.5	1	yes	2.496	3.733	0.9993	0.2611	499.3	1.061	0.7154	0.5351
2.5	10^{-1}	yes	2.499	3.606	0.9845	0.2497	500.4	1.063	0.7555	0.5463
2.5	10^{-2}	yes	2.5	3.18	1.05	0.3023	500.7	1.063	0.7595	0.583
2.5	10^{-3}	yes	2.5	3.169	1.046	0.2989	500.7	1.063	0.7599	0.5811
3.0	1	yes	2.984	4.526	0.9727	0.2201	543.2	1.056	0.7154	0.6163
3.1	1	yes	3.085	4.601	0.9764	0.2189	551.8	1.055	0.7154	0.636
3.2	1	yes	3.184	4.226	0.9883	0.2234	560.1	1.054	0.7154	0.6641
3.3	1	yes	3.284	4.704	0.9921	0.2225	568.3	1.053	0.7154	0.685
3.4	1	yes	3.385	4.563	0.9932	0.2198	576.5	1.052	0.7154	0.7055
3.5	1	yes	3.491	4.572	1.112	0.3039	584.8	1.051	0.7154	0.8112
4.0	1	yes	4	10.36	0.9986	0.2054	624.1	1.048	0.7154	0.8313
5.0	1	yes	5	12.13	0.9695	0.168	694.2	1.042	0.7154	0.994

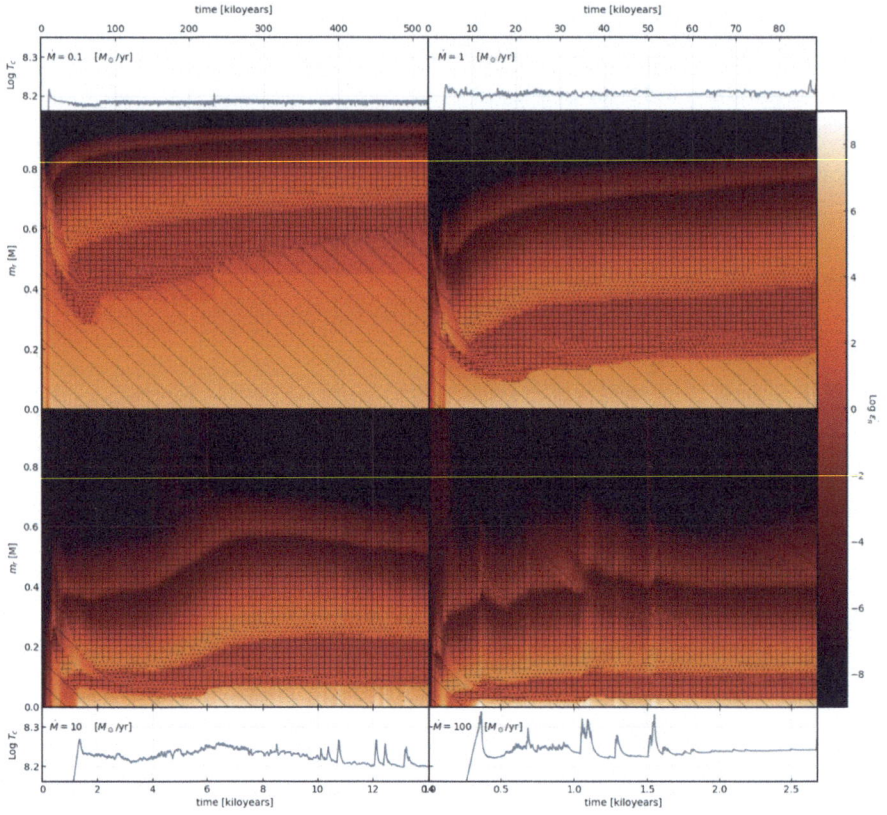

Fig. 3.5 Main panels—Kippenhahn diagrams for the Pop III models showing each decade in accretion rate. The axes are time and mass coordinate divided by total mass. The color shows the logarithmic energy generation rate, while the hatches show convective (diagonal), semi-convective (crossed) and radiative (dotted) regions. Accompanying panels—Central temperature time evolution as well as labels for each accretion rate

higher accretion rates. This may depend on the complex interactions of convection, energy generation, and entropy, or it may just be that the convective cores grow over time, and the higher the accretion rate, the less time they have to grow. Indeed, the growth of the convective core does seem to correlate somewhat with temperature spikes, though the cause of these spikes is not immediately clear. The average temperature also appears to increase for higher accretion rates, as more energetic nuclear burning may be required to sustain the more rapidly accreting SMSs. In Fig. 3.5, two burning regions are visible outside of the core region. These correspond to burning of primordial ^{3}He and ^{6}Li, with the lithium region being the outermost burning region visible in each panel.

Table 3.2 shows the results of the accreting stellar evolution simulations. We find that the final mass of the star grows with accretion rate, though rates of 100 $M_\odot$/yr and

3.2 Hydrodynamics

Table 3.2 Summary table for the evolution of the accreting SMSs recording accretion rate, metallicity, central temperature, total mass, and core mass at the GR instability

$\dot{M}$ [M$_\odot$/yr]	Z [Z$_\odot$]	Log $T_{c,\text{GRI}}$	M_{GRI} [M$_\odot$]	$M_{\text{core,GRI}}$ [M$_\odot$]
0.1	0	8.184	5.325e4	2.3754
1	0	8.202	9.356e4	1.756e4
10	0	8.199	1.418e5	9349
50	0	8.226	2.138e5	5881
90	0	8.239	2.648e5	1.02e4
100	0	8.253	2.742e5	7148
200	0	8.254	1.846e5	6954
1	10^{-3}	8.012	1.305e5	3.411e4
10	10^{-3}	8.055	1.791e5	2.204e4
100	10^{-3}	8.124	1.4e5	7364
1	10^{-1}	7.868	1.542e5	4.886e4
10	10^{-1}	7.886	2.484e5	2.25e4
100	10^{-1}	8.032	1.651e5	1.35e4
200	10^{-1}	8.013	1.409e5	1.013e4
1	1	7.808	2.986e5	1.188e5
10	1	7.805	2.076e5	4.294e4

up do not follow this trend. This finding is in slight tension with previous work [32] and we will investigate it further in future work. We also find that the GR instability occurs at lower temperatures for increasing metallicity, and this is due to the larger opacities for these models. As was mentioned above, the mass of the core roughly follows the lifetime of the star before the GR instability as a proportion of the total mass. Thus, low accretion rate models which survive for millions of years have the majority of their mass tied up in the convective core (Fig. 3.5).

3.2 Hydrodynamics

In this section, we review the results of the hydrodynamics simulations for each of the three cases. In each subsection, we will first describe the details of an example explosion, before turning towards the conditions and properties of the explosion for the larger population of SMSs.

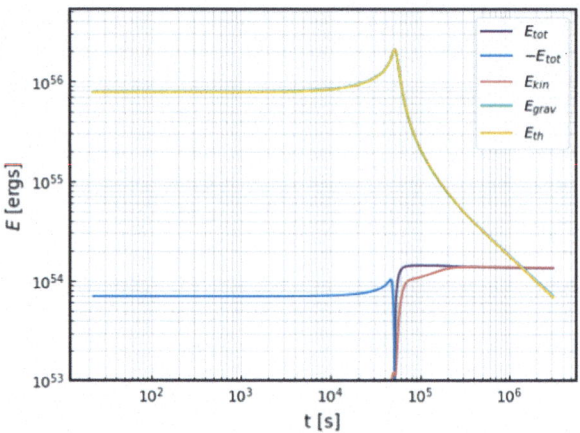

Fig. 3.6 Time evolution of energies during the hydrodynamics calculation (total, kinetic, gravitational, and internal/thermal, respectively) for the the 3 $\times 10^4$ $M_\odot$ SMS. Reprinted from Nagele et al. [21], with permission from MNRAS

3.2.1 Metal Free

GRSNe in primordial SMSs are driven by the explosive α process, and thus can be found in models which experience the GR instability during helium burning. For the explosion of 3×10^4 $M_\odot$—which we will use as an example in the figures—Fig. 3.6 displays the evolution during the hydrodynamics simulation of several energies, including the total energy which translates to the explosion energy at shock breakout (Table 3.3). There are three phases, the initial contracting phases with $E_{\text{tot}} < 0$, the pre-shock breakout phase with $E_{\text{tot}} > E_{\text{kin}} > 0$, and the post-shock breakout phase with $E_{\text{tot}} = E_{\text{kin}} > 0$. For both pulsating and exploding models, nuclear burning is limited to the inner region of the core, with the extent of the burning determined primarily by the maximum temperature which a particular mesh reaches during the contraction.

After the explosive nuclear burning, the inwards velocity rapidly reverses and the shock propagates towards the surface of the SMS (Fig. 3.7) with a typical velocity of a few percent the speed of light (Table 3.3); shock breakout occurs on a timescale of 10^5 s. The initial inwards velocity is largest in the envelope, and we emphasize that the GRSN involves the collapse of the entire star, not just the core. This is another reason why the stability analysis is necessary, as an analysis of the core alone will not always capture the instability. In the exploding case (Fig. 3.7, left), the final velocity is monotonically increasing, while in the pulsating case (Fig. 3.7, central) the final velocity nears zero in most of the star, is slightly negative at the edge of the remnant, before rapidly increasing with the ejected material. In the bottom row of this figure, the corresponding discontinuity in radius is clearly visible.

For the non accreting metal free stars, we find that most of the models collapse, but there are several pulsations and two GRSNe (Table 3.3). The main difference between the pulsations and the explosions is that the explosions are significantly more energetic, with the pulsations only producing enough energy to unbind a small fraction of the envelopes. As we will show later, the pulsations are, in fact, the more interesting of the two types when it comes to detecting these events by their lightcurves. Thus, it is worth taking a moment to comment on why previous studies did not find pulsations, despite identifying GRSNe. The crux of the matter, as with several other topics in this thesis, has to do with the use of the GR stability analysis. Nagele et al. [20]; Chen et al. [6] began the hydrodynamics calculations when the SMSs were already extremely unstable. Thus, models which we found in this work to pulsate, would instead collapse to black holes because the initial conditions for the hydrodynamics were too late in the evolution of the star. Thus, the use of the stability analysis allows a larger variation of behavior to be captured, incuding both the extension of the mass window for GRSNe and the existence of the thermonuclear pulsations, which is analogous in some ways to the pulsational pair instability [33].

For the pulsations, we determine the ejecta mass in Table 3.3 using the local energy, e(r) (Fig. 3.8, left panel), which is the integrand of the global energies defined in Sect. 2.1. We measure this quantity at shock breakout (right panel), when some of the energy is still in the form of thermal energy. The energy evolution after shock breakout may not be completely accurate because the energy is evolved using the entropy equation, and energy conservation is not guaranteed in extremely low density regions where the accuracy of the EOS suffers. We confirm that the escape velocity criterion (right panel) converges roughly to this same value, validating the use of this ejection criterion. Note that 2.9×10^4 $M_\odot$ is a marginal case. Although it is not an explosion, it does not have the steady behaviour of Fig. 3.8 after shock breakout, and we expect the value in Table 3.4 to underestimate the ejecta mass for this case.

The mass range for the explosion follows from straightforward considerations. Models which experience the instability before helium burning has reduced the binding energy of the star cannot explode. On the other hand, models with helium mass fractions less than ten percent also do not explode or pulsate because they lack fuel for alpha capture reactions sufficient to halt the collapse. The pulsating models all have lower mass than the exploding models and the most massive model being the most energetic before a sharp drop-off is reminiscent of Fig. 6 of Nagele et al. [20]. This is likely a characteristic of GRSNe even if the mass range found by the current analysis is not completely correct.

Figure 3.9 shows the behavior of the central temperature, baryonic density, nuclear energy generation rate, and entropy as a function of time. For the exploding and pulsating models (left column), temperature and density increase and decrease smoothly, as was the case for the GRSNe in Chen et al. [6]; Nagele et al. [20], and the nuclear energy generation rate is also smooth. The models which collapse (right column) also show relatively smooth increases in temperature and density, but the energy generation rate and change in entropy are more complicated, as the stars enter different phases of burning and photodisintegration. Roughly speaking, the first three peaks in $\dot{\epsilon}_c$ (first peak at 10^9 K) correspond to carbon burning, sulfur burning, and calcium to iron burning. Stars with a more evolved core, such as 2.4×10^4 $M_\odot$ do not have much carbon remaining, and the first peak corresponds to neon burning. At the other extreme, stars with reserves of hydrogen, such as 4×10^4 $M_\odot$ do not exhibit peaks because the presence of protons increases the number of possible reactions. After the core becomes iron/nickel, the next major reactions are assisted by free nucleons

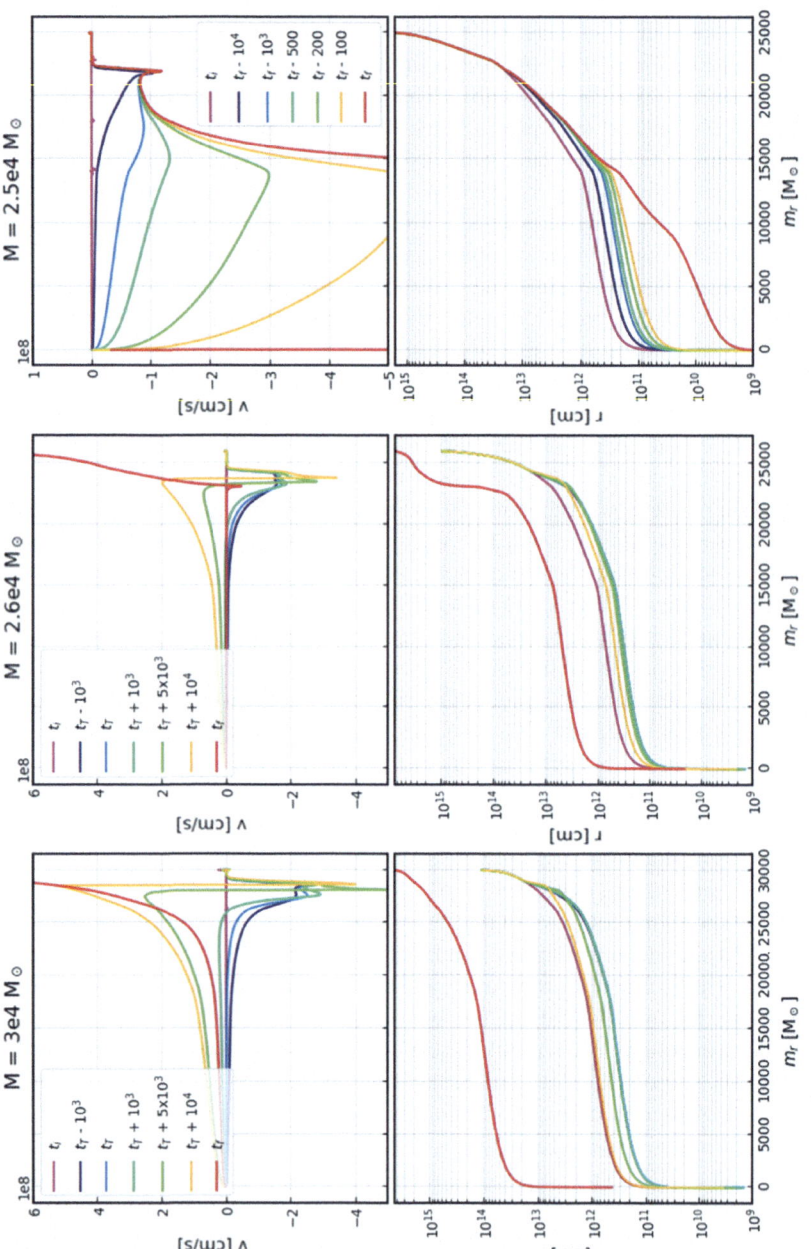

Fig. 3.7 Time snapshots of the velocity (upper panel) and radius (lower panel) as a function of mass for three models, an explosion, a pulsation, and a collapse, respectively. Reprinted from Nagele et al. [21], with permission from MNRAS

3.2 Hydrodynamics

Table 3.3 Summary table for all metal free models. The columns are total mass, outcome of the hydrodynamics code, mass of the isentropic core, central helium mass fraction at the start of the hydrodynamics code, change in helium mass fraction, explosion energy, maximum central temperature, and maximum velocity of the outermost mesh point, denoted v_R

M [10^4 $M_\odot$]	Outcome	M_{core} [$M_\odot$]	$X_c(^4He)$	$\Delta X_c(^4He)$	E_{exp} [ergs]	max T_c [K]	max v_R/c
2	Collapse	10926	1.37e−3	–	–	–	–
2.1	Collapse	11368	2.23e−4	–	–	–	–
2.2	Collapse	11729	1.22e−4	–	–	–	–
2.3	Collapse	12595	3.17e−2	–	–	–	–
2.4	Collapse	13180	3.44e−18	–	–	–	–
2.5	Collapse	13798	2.69e−3	–	–	–	–
2.6	Pulsation	14772	0.104	0.104	4.32e53	7.58e8	0.032
2.7	Pulsation	14964	0.222	0.147	4.70e52	6.62e8	0.021
2.8	Collapse	15596	0.713	–	–	–	–
2.9	Pulsation	16183	0.589	0.153	7.56e53	7.33e8	0.041
2.95	Explosion	16504	0.599	0.168	1.23e54	7.69e8	0.046
3	Explosion	16817	0.652	0.152	1.43e54	8.06e8	0.048
3.05	Collapse	17144	0.734	–	–	–	–
3.1	Collapse	17516	0.794	–	–	–	–
3.15	Collapse	17793	0.815	–	–	–	–
3.2	Collapse	18091	0.815	–	–	–	–
3.3	Collapse	18888	1.000	–	–	–	–
3.4	Collapse	19460	1.000	–	–	–	–
3.5	Collapse	19933	0.950	–	–	–	–
4	Collapse	23891	0.960	–	–	–	–

Table 3.4 Mass ejecta by isotope for the explosions and the pulsations. Except for the first column which is consistent with Table 3.3, values are recorded in units of $M_\odot$. Yield tables for the explosions are available online

M [10^4 $M_\odot$]	M_{ej}	M(^{1}H)	M(^{4}He)	M(^{12}C)	M(^{16}O)	M(^{20}Ne)	M(^{24}Mg)	M(^{28}Si)	M(^{32}S)
2.6	2808	1877	974	<0.1	<0.1	<0.1	<0.1	<0.1	<0.1
2.7	2299	1584	759	<0.1	<0.1	<0.1	<0.1	<0.1	<0.1
2.9	2078	1465	651	<0.1	<0.1	<0.1	<0.1	<0.1	<0.1
2.95	29500	5441	16946	3006	2505	481	812	306	1.2
3	30000	5537	18077	2986	1829	367	702	497	5.1

created by photodisintegration (one peak), which is finally followed by photodisintegration of nickel (one peak) and photodisintegration of helium (two peaks). In reality, neutrino reaction would then begin to dominate, but the hydrodynamics code does not include all of the relevant neutrino reactions.

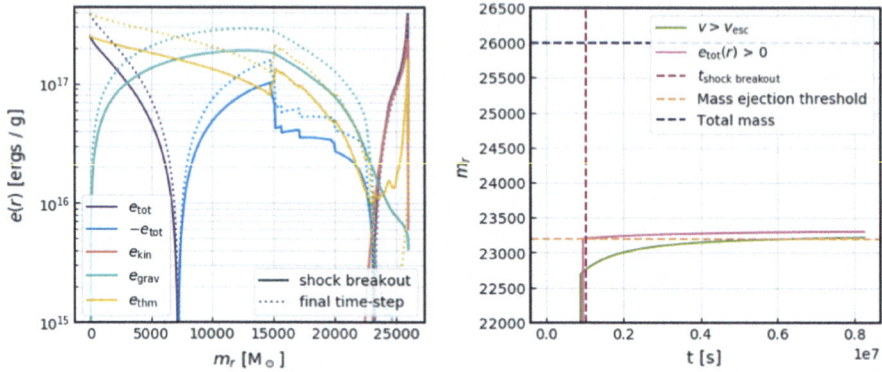

Fig. 3.8 Left panel—local energy as a function of mass coordinate for the 2.6×10^4 $M_\odot$ metal free model (pulsation). The ejection criterion is $e_{tot} > 0$ for all $m_r > M - M_{ej}$ at the time of shock breakout. Right panel—illustration of the stability of the ejection criterion. We measure the ejected mass at shock breakout and this quantity is roughly consistent with the post Newtonian escape velocity criterion near the end of the simulation. Reprinted from Nagele et al. [21], with permission from MNRAS

Having discussed the explosion and collapse, we now turn to the eventual fate of the pulsating models, and whether or not they will pulse again [19]. We will discuss the properties of the second pulsation for each of the three pulsating models in turn. The lowest mass model (26000 $M_\odot$) pulsates (at least) twice, with the second pulse having lower explosion energy and ejected mass than the first pulse. The energy produced via nuclear burning is also smaller in the second pulse, though the difference is moderate.

The intermediate mass model (27000 $M_\odot$) also experiences a second pulsation, but it is not strong enough to eject mass according to the criterion of Nagele et al. [21]. This is because the maximum temperature is significantly lower than the other models (Fig. 3.9), resulting in less energy generation due to nuclear burning. We note that the final velocity structure suggests a small amount of mass ejection ($\sim$ 100 $M_\odot$), and thus we will need to reexamine our mass ejection criterion in future work.

The highest mass model (29000 $M_\odot$) does not experience a second pulsation and instead collapses to a black hole. However, we have confirmed that by artificially increasing the oxygen mass fraction by a small amount (0.17–0.2), we can induce an explosion which is a factor of three times more energetic than those found in Nagele et al. [21]. Because the lower mass models have high oxygen mass fractions, it is likely that in between the intermediate mass model and the high mass model, there exists pulsating models which explode during the second pulse. We plan to investigate this phenomenon in future work.

Although we only analyze three pulsating models, hints of a trend are present. The lower and intermediate mass models have plenty of oxygen, but low levels of helium. In contrast, the higher mass model has plenty of helium, but low oxygen. This

3.2 Hydrodynamics

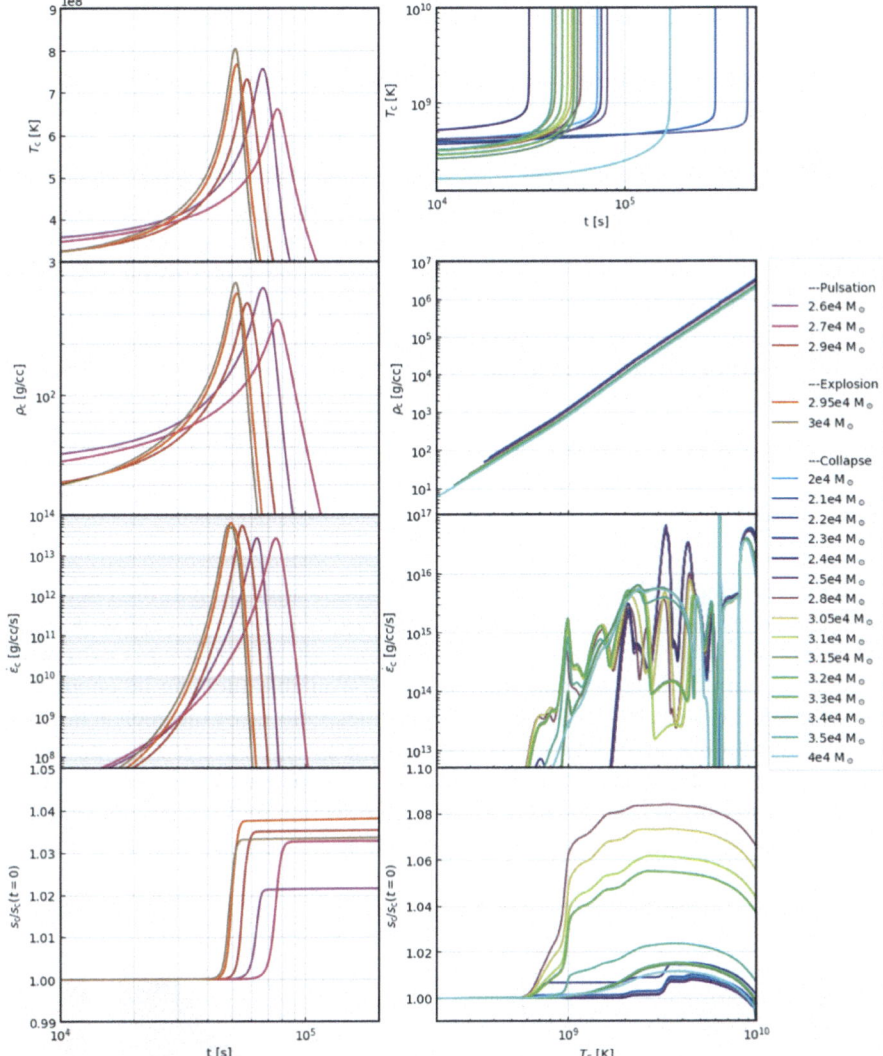

Fig. 3.9 Results of the hydrodynamics calculation for metal free GR unstable HOSHI models. Time evolution of central temperature (1st panel), density (2nd panel), rate of change of specific internal energy (3rd panel) and entropy relative to the initial value (4th panel). The legend groups models by outcome, whereas colors vary with mass. The left column shows the exploding and pulsating models as a function of time. The right column shows the temperature as a function of time, while the other three panels are functions of temperature. Reprinted from Nagele et al. [21], with permission from MNRAS

pattern suggests that somewhere in between these models, there may be a favorable regime for strong pulsations, where the stars have large reservoirs of both oxygen and helium. Note that the oxygen reservoir is necessary because $^{12}C(\alpha, \gamma)^{16}O$ is subdominant in this temperature regime so that the explosive alpha process begins with ^{16}O$(\alpha, \gamma)^{20}$Ne.

There are three requirements for a pulsating model to pulsate again. (1) The star must be stable to the GR radial instability while it contracts yet unstable to convection. (2) The GR instability should occur soon after the end of the contraction. If condition (1) is not satisfied, then the star will not be able to remix and transport fresh nuclear fuel to the center. If condition (2) is not satisfied, the fresh nuclear fuel which has been transported to the center is liable to be consumed by burning on evolutionary timescales. (3) The helium and oxygen abundances in the center of the star must not be too low at the GR instability. Conditions (1) and (2) are checked numerically and contribute to condition (3) being satisfied. However, for the 29000 $M_\odot$ model, condition (3) is not satisfied even though the other two are, and this model collapses to a black hole after the second instability.

3.2.2 Metal Rich

We now turn to the explosions of metal rich SMSs. If a metal rich SMS becomes unstable in the hydrogen burning phase, the only requirement for an explosion is the existence of sufficient seed nuclei. This is because hydrogen burning is extremely energetic, so burning even a small fraction of the available fuel can be enough to unbind the star. Figure 3.10 shows the velocity profiles, colored by the nuclear heating rate, for several time snapshots around the maximum temperature. Time proceeds from the bottom of the figure to the top, as the velocity increases. Figure 3.10 shows that the explosive nuclear burning occurs in a large section of the star, with stronger heating nearer the center. Nuclear reactions continue to occur until the temperature becomes sufficiently low.

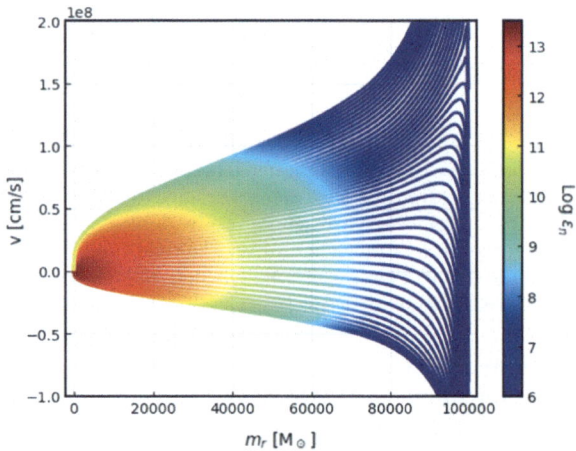

Fig. 3.10 Velocity profiles during the hydrodynamics calculation for a metal rich GR unstable HOSHI model (3.4×10^5 $M_\odot$). Color denotes log heating due to nuclear reactions. Reprinted from Nagele et al. [18], with permission from MNRAS

3.2 Hydrodynamics

The nucleosynthesis at work during the explosion can be broadly separated into two categories. First, there are cyclical processes such as the hot CNO process which are more prevalent towards the bottom left of Fig. 3.11. In addition, there is the low temperature rp process whereby possibly multiple proton captures are followed by inverse beta decays, which are then followed by more proton captures. The key ingredient in this process is the presence of free protons, a condition which is obviously satisfied if the GR instability occurs during the hydrogen burning phase. The balance between these two types of nucleosynthetic processes is determined by the maximum temperatures, with lower temperatures being associated more with cyclical processes. It is important to note that since the maximum temperature decreases smoothly as one travels outwards in mass coordinate, that this balance also changes in different regions of the core.

The explosion energy and maximum outflow velocities of all models are summarized in Table 3.5. Figure 3.12 shows explosion energy (triangles) as a function of initial mass and metallicity for mass loss (left panel) and no mass loss (right panel). The inclusion of mass loss primarily effects the timing of the GR instability, but the explodibility and explosion energy are determined by the mass and metallicity, and thus these largely do not depend on whether or not mass loss is included, although the one exception to this trend is the solar metallicity models. These models sometimes lose enough mass so that they become small enough that they do not experience the GR instability. As is apparent, more massive models require higher metallicity (more seed metals for proton captures) in order to explode.

In order to estimate a condition for explosion, we compute the amount of energy available due to nuclear reactions in a star with mass M, assuming that the rp process proceeds up to an isotope with proton number m (so we are interested in nuclear potential energy, $P_{\text{nuc},m}(M)$). We assume that every element with atomic number greater than 15 is converted to the most abundant solar isotope with proton number m (I_m). Then, the chemical distribution of the star will change as follows:

$$X'_{I_j} = \begin{cases} X_{I_j} & A(I_j) < 16 \text{ or } A(I_j) > A(I_m) \\ 0 & 15 < A(I_j) < A(I_m) \end{cases} \quad (3.4)$$

$$X'_p = X_p - \sum_{I_j} (X_{I_j} - X'_{I_j}) \frac{A(I_m) - A(I_j)}{A(I_j)} \quad (3.5)$$

$$X'_{I_m} = X_{I_m} + X'_p + \sum_{I_j} (X_{I_j} - X'_{I_j}) \quad (3.6)$$

where unprimed values are the initial mass fractions and primed values are the final ones. $P_{\text{nuc},m}(M)$ is the energy released by this change in composition. Then, by comparing this energy to the star's gravitational energy, we estimate that a phenomenological condition for explosion in the 10^5 M$_\odot$ models is $P_{\text{nuc},m}(10^5 \text{ M}_\odot) > 0.15 E_{\text{grav}}$. This exercise allows us to test the coverage of the 153 isotope network which has

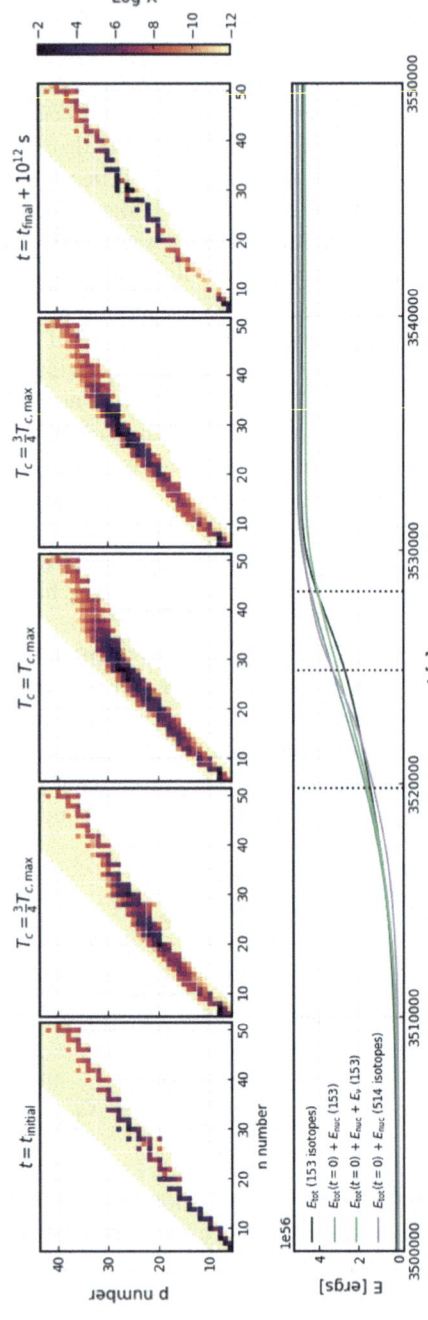

Fig. 3.11 Nucleosynthesis in the post processed 514 isotope network for the $M = 3.4 \times 10^5 \, M_\odot$, $Z = Z_\odot$ model. Upper panels—isotope mass fractions at the central mesh for five snapshots: the initial time, when the temperature rises to 3/4 of the eventual maximum, the maximum temperature, when the temperature falls to 3/4 of the maximum, and 10^{12} seconds after the end of the hydrodynamical calculation. Lower panel—total energy compared to E_{nuc} in both the 153 and 514 isotope calculations and compared to $E_{nuc} + E_\nu$ for the 153 isotope calculation. The vertical dashed lines show the times of the central three panels. Reprinted from Nagele et al. [18], with permission from MNRAS

3.2 Hydrodynamics

Table 3.5 Summary of the explosions in the hydrodynamical calculation and post processing for the metal rich models. The columns are, initial mass, metallicity, and mass loss, followed by maximum central temperature, the change in central hydrogen mass fraction, explosion energy, maximum outflow velocity, and the ejected mass of ^{14}N, ^{45}Sc, and ^{56}Ni in units of $M_\odot$

M [10^5 $M_\odot$]	$Z/Z_\odot$	$\dot{M}$	$T_{c,max}$ [10^7 K]	$\Delta X_c(^1H)$	E_{exp} [10^{54} ergs]	max(v) [10^8 cm/s]	M(^{14}N)	M(^{45}Sc)	M(^{56}Ni)
1.0	1	no	21.15	0.07972	7.345	8.208	331.9	0.003939	2.687e−15
1.0	10^{-1}	no	36.54	0.03783	8.493	10.92	80.03	0.01034	2.501e−7
1.0	10^{-1}	yes	33.71	0.03388	7.971	9.764	79.53	0.001718	1.875e−8
1.0	6×10^{-2}	yes	28.41	0.02573	4.312	8.517	64.67	0.0004773	4.646e−11
1.0	8×10^{-2}	yes	36.42	0.03127	7.391	10.06	545.5	0.2503	5.965e−6
1.1	10^{-1}	yes	41.83	0.03798	11.41	11.63	79.76	0.2495	2.818e−5
1.5	1	no	27.26	0.1574	42.03	13.25	1203	0.006465	1.14e−11
1.5	1	yes	27.3	0.1213	21.68	10.12	833.4	0.004427	1.675e−12
2.0	1	no	27.34	0.1778	64.62	11.41	866.4	0.009184	4.821e−11
2.5	1	no	30.3	0.2007	107.1	13.57	1165	0.01735	2.732e−9
2.5	1	yes	23.43	0.1775	57.52	9.177	932.9	0.009946	1.685e−13
3.0	1	yes	34.01	0.2204	162	15.73	1476	0.04457	2.235e−7
3.1	1	yes	37.09	0.2239	186.7	16.89	1573	0.1745	5.295e−6
3.2	1	yes	41.44	0.2284	213.5	18.23	1669	4.664	0.0002803
3.3	1	yes	47.15	0.2359	244.6	19.67	1745	18.85	0.0222
3.4	1	yes	59.29	0.2511	298.3	22.07	1687	25.93	180.7

maximum proton number $m = 30$. If we had the computational resources to couple the 514 isotope network, this would instead extend to $m = 42$. We then check whether any of the collapsing 10^5 $M_\odot$ models satisfy the above condition for $m = 42$. The $Z = 0.04$ $Z_\odot$ model satisfies the condition, but the $Z = 0.02$ $Z_\odot$ model does not and we hypothesize that extending the nuclear network would not drastically reduce the metallicity threshold for the explosion.

3.2.3 Accreting

We now turn to the explosions and pulsations of accreting supermassive stars. It had previously been unknown whether accreting SMSs could explode, but we show here several example of explosions and pulsations (Table 3.6). The consideration at play is that accreting SMSs are much less compact than their non accreting cousins, and because of this, energy injection due to nuclear burning has longer to arrest the collapse of the star. In the Pop III case, this often results in a pulsation, ejecting a large quantity of mass which is likely observable as a transient. In the metal rich case, not

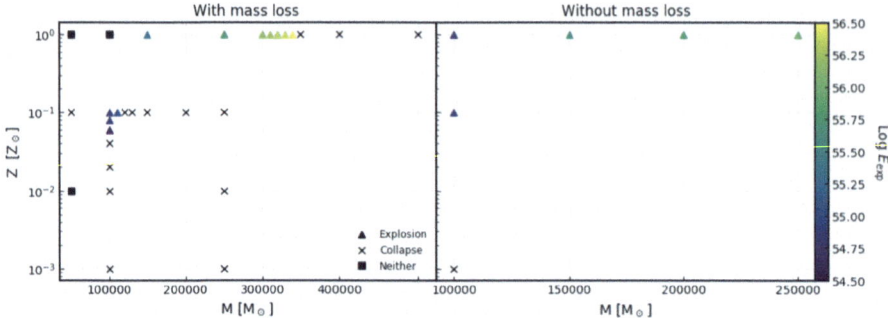

Fig. 3.12 Results of the hydrodynamics calculation for metal rich GR unstable HOSHI models. Dependence of explosion energy (color) for the exploding models (triangles) on mass and metallicity. The black crosses are models which failed to explode and black squares either do not reach the GR instability (because of large mass loss) or are stable. The left panel shows models with mass loss while the right panel shows those without. Reprinted from Nagele et al. [18], with permission from MNRAS

Table 3.6 Summary table for the hydrodynamical simulations of the accreting SMSs recording accretion rate, metallicity, the outcome of the hydrodynamical simulation, maximum temperature, nuclear energy generated, mass ejected and kinetic energy of the ejected material during the hydrodynamical simulation. Reprinted from Nagele et al. [17], with permission from PhRvD

$\dot{M}$ [M$_\odot$/yr]	Z [Z$_\odot$]	Outcome	Log $T_{c,max}$	E_{nuc} [10^{54} ergs]	$M_{ejected}$ [M$_\odot$]	E_{kin} [10^{54} ergs]
0.1	0	Collapse	–	–	–	–
1	0	Pulsation	8.709	1.059	3051	0.1312
10	0	Pulsation	8.471	0.6154	0	0
50	0	Pulsation	8.519	1.138	1046	0.01036
90	0	Collapse	–	–	–	–
100	0	Collapse	–	–	–	–
200	0	Pulsation	8.649	1.792	3878	0.1255
1	10^{-3}	Collapse	–	–	–	–
10	10^{-3}	Pulsation	8.343	0.88	0	0
100	10^{-3}	Pulsation	8.47	0.8506	0	0
1	10^{-1}	Pulsation	8.236	1.27	4312	0.1733
10	10^{-1}	Pulsation	8.176	0.9747	0	0
100	10^{-1}	Explosion	8.297	4.725	1.651e5	1.043
200	10^{-1}	Explosion	8.28	3.196	1.409e5	0.3221
1	1	Explosion	8.274	17.26	2.986e5	10.3
10	1	Pulsation	8.069	0.8976	0	0

only will there be an electromagnetic transient, but also interesting nucleosynthetic yields as the nuclear burning resembles that of the previous section (see Sect. 3.4.2).

3.3 Supernova Lightcurves

In the following section, we will discuss modeling the lightcurves associated with shock breakout for the metal free explosions. We have not yet applied these techniques to the metal rich explosions, but given the more energetic and longer nature of these events, they would seem to be even more exceptional than the models discussed here. Their investigation, however, will be left for future work.

3.3.1 Metal Free

For the metal free models, the peak luminosity occurs soon after shock breakout, and is determined primarily by the photosphere radius at shock breakout. This means that the pulsations of the low mass models radiate very strongly, despite their small explosion energies. For the first day (rest frame), the luminosity decreases monotonically as does the effective temperature. At around one day (or around four days for 26000 $M_\odot$, first pulsation), helium ions begin recombining and this leads to a slight uptick in the luminosity. Once the He ionization fraction has dropped, the luminosity continues to decrease until around twenty days, when hydrogen ions begin to recombine (visible in Fig. 3.13). This is the beginning of the plateau phased identified in Moriya et al. [14]. The luminosity increases until the photosphere recession velocity matches the hydrodynamic expansion, at which point the photosphere stalls, causing the luminosity to plateau [14].

We find that the duration of this plateau phase in our models is roughly half that found by Moriya et al. [14] (Fig. 3.13). There are several possible explanations for this discrepancy. The first is that due to the earlier collapse of our models, due primarily to our utilisation of the GR stability analysis, the envelope composition of the progenitors is different, causing a difference in recombination. Next, it is possible that differences between SNEC and STELLA, in particular multi-wavelength transfer and energy loss from the photosphere, impact the timing of the plateau phase. A more plausible explanation is that the exploding model used in Moriya et al. [14] is actually the 2D model of Chen et al. [6] which is nearly an order of magnitude more energetic than our explosions, with the extra energy produced in part by Kelvin Helmholtz instabilities. It is worth reiterating at this point that our brightest model are actually the puslating ones, which although they have lower total energies, have much larger progenitor radii, leading to comparable luminosities to the extremely energetic model in Moriya et al. [14].

We calculate the AB magnitude of our models in the reddest band of each of JWST, Euclid, Roman, and Rubin (Fig. 3.13). As in Moriya et al. [14], we assume ΛCDM cosmology with $H_0 = 70 \, \text{km s}^{-1} \, \text{Mpc}^{-1}$, $\Omega_M = 0.3$. JWST bands and cosmological distances were both calculated using astropy. We find that the plateau phase (100 days rest frame) of the second pulsation for 26000 $M_\odot$ is visible to JWST at $z > 12$. Also of interest is plateau phase of the first pulsation, which is visible up to $z \sim 18$.

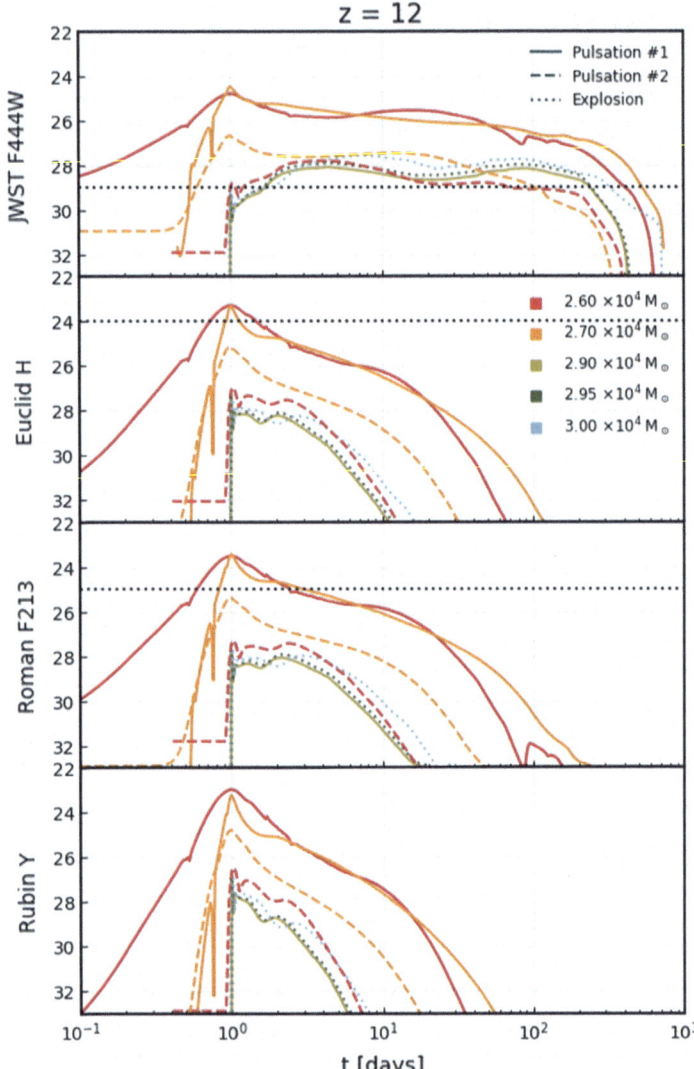

Fig. 3.13 Results of SNEC simulations for metal free exploding and pulsating models. The four panels show AB magnitudes for GRSNe occurring at redshift 12 for JWST, Euclid, Roman and Rubin, and specificalyy the reddest band of each instrument. Colors denote masses and solid lines show the furst pulsation, while dashed lines show the second and dotted lines show exploding models for comparison. The dotted horizontal lines show a typical limiting magnitude for each of the first three instruments. Reprinted from Nagele et al. [19], with permission from MNRAS

3.4 Nucleosynthetic Yields

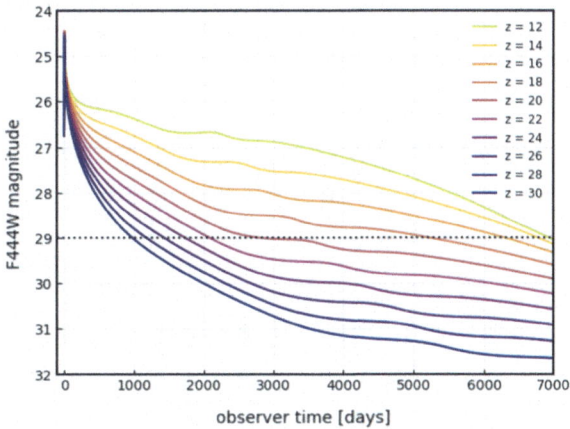

Fig. 3.14 Same as Fig. 3.13, but only for the most luminous model observed by JWST with the colors showing redshift dependence. Reprinted from Nagele et al. [19], with permission from MNRAS

At $z = 30$, the first pulsation of 27000 $M_\odot$ is visible for more than 1000 days in the observer frame (Fig. 3.14). It is worth mentioning that because of the long duration of these events, identifying them as transients in the first place may be difficult. This problem is ameliorated either if the event is caught near the peak instead of during the plateau or if multi-band information is available. In the latter case, Moriya et al. [14] showed that GRSNe can be differentiated from other persistent sources using only three JWST bands.

3.4 Nucleosynthetic Yields

3.4.1 Metal Free

For the metal free GRSNe, the explosion yields are straightforward. They include α process elements, particularly magnesium and silicon, up until a certain mass which the α process does not reach for the low temperatures at which the GRSNe occur (calcium or argon). This results in a sawtooth pattern similar to the yields of pair instability supernovae. Figure 3.15 shows these yields for the two explosions, and comparisons with metal poor stars, none of which can plausibly reproduce the abundance pattern.

3.4.2 Metal Rich

The metal rich explosions have two nucleosynthetic signatures. First is enhanced nitrogen relative to the solar abundance [1] and to carbon and oxygen. This is essen-

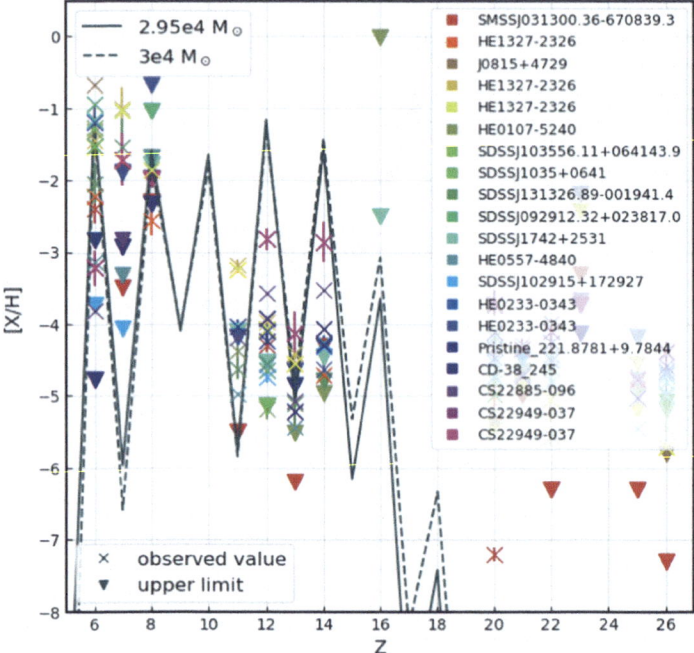

Fig. 3.15 Metal free explosion yields (lines) compared with observed metal poor stars (crosses are observations, triangles upper limits). The line shows the fraction relative to a minimum mass of hydrogen with which it would have to mix Magg et al. [12], and can thus be regarded as an upper limit. Reprinted from Nagele et al. [21], with permission from MNRAS

tially due to the CNO cycle producing order of magnitude equal amounts of its eponymous elements. In addition, the cyclical and low temperature rp processes means that light elements are exchanged for slightly heavier ones. This manifests in a dearth of fluorine specifically, with the final abundances being rich in chlorine, potassium, scandium and vanadium (Fig. 3.16). The three panels in this figure show models without mass loss, those with mass loss at solar metallicity, and those with initial mass of 10^5 $M_\odot$, respectively. In general, models with more distinctive yield patterns reach higher maximum temperatures.

In Fig. 3.17, we show the spatial variation of yields from three models. This is due to the fact that each explosion is producing thousands of solar masses of nucleosynthetic material, which varies strongly with the mass coordinate. We perform this demonstration for a fiducial model ($M = 10^5$ $M_\odot$, $Z = Z_\odot$, without mass loss), a metal poor model ($M = 10^5$ $M_\odot$, $Z = 0.08$ $Z_\odot$, with mass loss) and a massive model ($M = 3.4 \times 10^5$ $M_\odot$, $Z = Z_\odot$, with mass loss). Also shown (right panels) are the velocity profiles near the end of the hydrodynamics simulation. In the fiducial model, nitrogen enhancement is only seen in the inner half of the star, whereas for the other two models, it extends to about 80% of the star. In the central 10% of the massive model, cobalt is suppressed because its lightest stable isotope (^{59}Co) cannot

3.4 Nucleosynthetic Yields

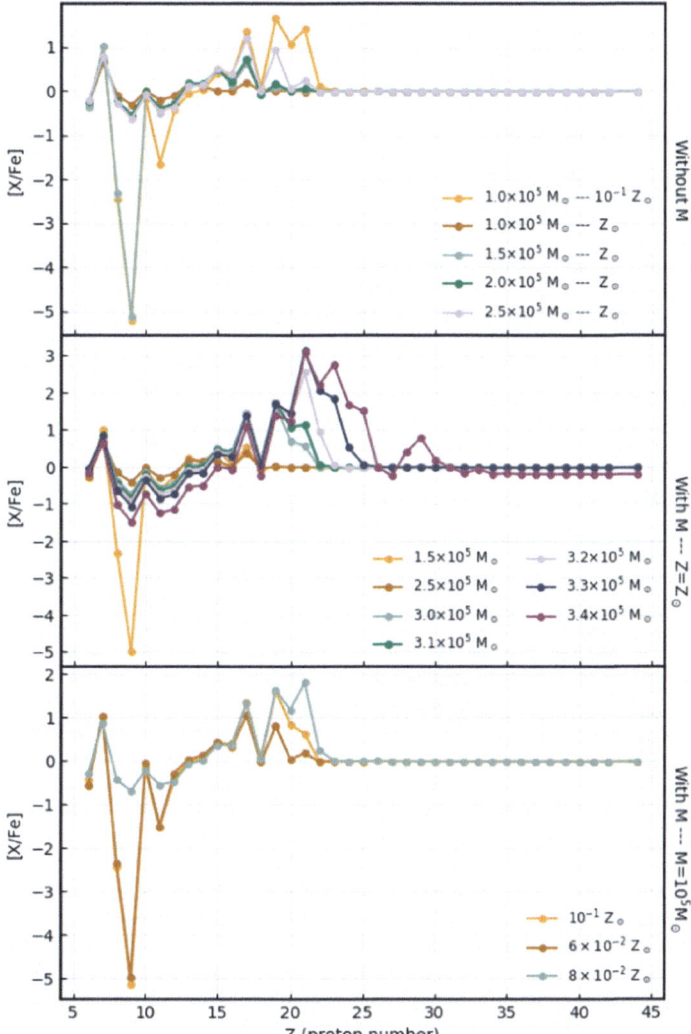

Fig. 3.16 Elemental yields of the explosion from the 514 isotope network for metal rich models. The panels show models without mass loss (upper panel), solar metallicity models with mass loss (central panel) and $M = 10^5$ M$_\odot$ models with mass loss (lower panel). Reprinted from Nagele et al. [18], with permission from MNRAS

be reached from the p side. Also in this region, ^{56}Fe is synthesized (Table 3.5) which shifts the heavy mass elements to sub-solar abundances (absolute abundances do not change).

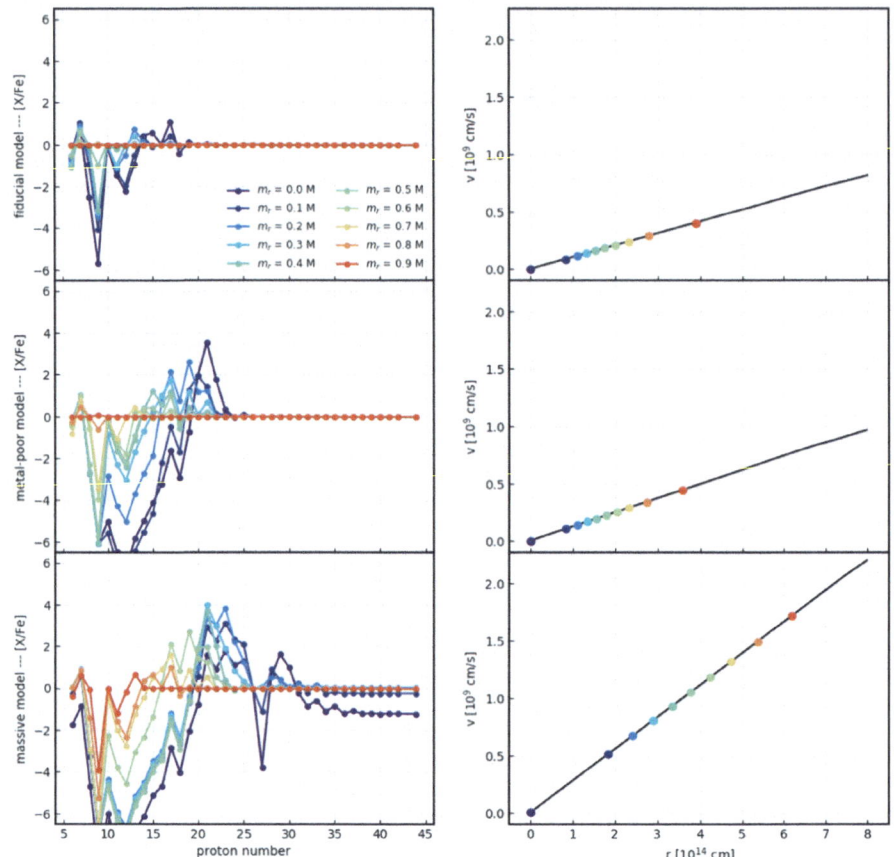

Fig. 3.17 Elemental yields of the explosion at selected mass coordinates, and velocity profiles for the three indicated metal rich models. Left column—elemental abundance for mass coordinates (not the average between mass coordinates) at multiples of 0.1 times the total mass. Right column—velocity profiles at a comparable time-step near the end of the hydrodynamics simulation for each of the three models. The velocity profiles are nearly homologous. Colored circles indicate the mass coordinate shown in the left column. Reprinted from Nagele et al. [18], with permission from MNRAS

3.5 Comparison to Super-Solar Nitrogen in GN-Z11

GN-z11 is a luminous galaxy slightly under redshift eleven with metallicity about ten percent the solar value [10, 22, 26]. GN-z11 was recently observed by JWST NIR-Spec to have unusually bright nitrogen lines, while simultaneously lacking the N V line (E > 77.5 eV) associated with an active galactic nucleus (AGN) [2]. Cameron et al. [3] analyze the NIRSpec observations and infer a N/O ratio two to four times the solar value, though they also conclude that an AGN origin cannot be ruled out. They propose several explanations for this enhanced nitrogen, including mass loss from

3.5 Comparison to Super-Solar Nitrogen in GN-Z11

rotating WN stars, fine tuned metal poor supernovae (SNe) yields, tidal disruption by an intermediate mass black hole, or winds from well mixed V/SMSs. Senchyna et al. [24] also analyze the NIRSpec observations and arrive at a similar conclusion regarding N/O. They then analyze spectra from nearby galaxies and suggest that GN-z11 could be a proto-globular cluster. Charbonnel et al. [5] compare the CNO abundances in GN-z11 to nearby globular clusters, and further expand on the suggestion that the origin of the enhanced nitrogen could be from V/SMSs formed via stellar collisions.

We now compare the four $Z = 0.1\, Z_\odot$ models to the observations of GN-z11. Super-solar oxygen is not seen in GN-z11 so it is worth pausing to make a general comment about wind compositions. SMS winds in the central hydrogen burning phase will be enriched by the CNO cycle (both central burning and shell burning), so that they contain super-solar nitrogen and sub-solar carbon and oxygen. However, during the central helium burning phase, carbon and oxygen begin to accumulate in the core, some of which is dredged up to the envelope. During this phase, the wind consists of super-solar nitrogen and roughly solar carbon and oxygen abundances. Finally, after helium burning, dredge up from the CO core continues and the wind becomes super-solar in C and O. While the exact details of the cumulative wind composition will depend on mass, metallicity and rotation, the end of life burning phase can be used as a guidepost.

After the $5 \times 10^4\, M_\odot$ and $10^5\, M_\odot$ models become GR unstable, we port the unstable stellar profiles into the hydrodynamics code to determine if they will explode or collapse. $5 \times 10^4\, M_\odot$ becomes unstable just at the end of hydrogen burning, and therefore it does not have enough hydrogen to power a CNO driven thermonuclear explosion [8, 13, 18] nor enough oxygen ($X \sim 10^{-4}$) to power an alpha process driven thermonuclear explosion [6, 20, 21] and therefore this model collapses. On the other hand, the $10^5\, M_\odot$ model has a healthy hydrogen reservoir and explodes with total energy 7.2×10^{54} ergs at shock breakout and a maximum outflow velocity of 9.8×10^8 cm/s.

Figure 3.18 shows the abundance ratios of nitrogen to oxygen, carbon to oxygen, and oxygen to hydrogen as in Fig. 1 of Cameron et al. [3] and Fig. 1 of Charbonnel et al. [5]. The grey regions are the conservative and fiducial constraints derived by Cameron et al. [3] (Table 1), where we have filled in the fiducial value of $12 + \log(O/H) \approx 8.1$ by inspection. The symbols show the yields from our models mixed with different masses of $Z = 0.1 Z_\odot$ gas as denoted by color. The yields from the lower mass models ($10^3, 10^4\, M_\odot$) are nitrogen rich, but also contain super-solar oxygen. These models survive into carbon-oxygen burning, at which point dredge up increases the oxygen and carbon abundances in the envelope. The implication of the oxygen rich winds is that although these two models produce super-solar nitrogen, they do not satisfy the fiducial constraints (dark grey region) and they only satisfy the conservative constraints (light grey region) for a narrow range of mixed ISM ($M_{ISM} \in (8e3, 9e4)\, M_\odot$ and $M_{ISM} \in (4e4, 5e5)\, M_\odot$ for the $10^3, 10^4\, M_\odot$ models, respectively).

The two higher mass models both end their lives before helium burning, thus producing super-solar nitrogen and sub-solar oxygen. The models both satisfy the

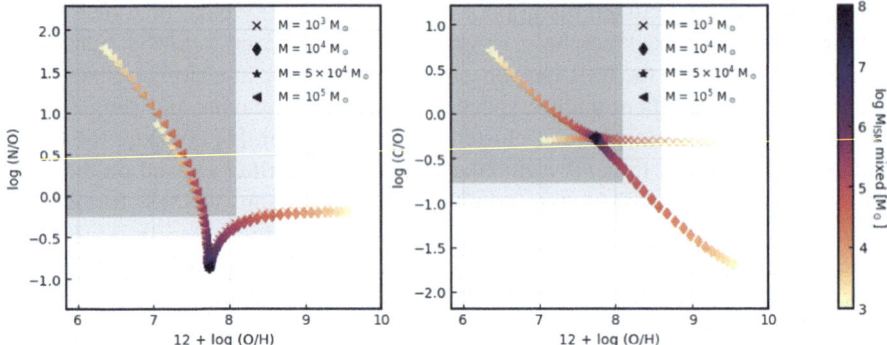

Fig. 3.18 Abundance ratios of nitrogen to oxygen, carbon to oxygen, and oxygen to hydrogen. The symbols denote each of the four models, while the color shows how much $Z = 0.1\,Z_\odot$ ISM the yields from each model have mixed with. The gray regions show the conservative (lighter) and fiducial (darker) constraints on GN-z11 from the model of Cameron et al. [3]. Reprinted from Nagele & Umeda [16] under the Creative Commons Attribution 4.0 licence

fiducial constraints for a wide mass range of mixed ISM ($M_{ISM} <$ 4e4 $M_\odot$ and $M_{ISM} <$ 5e5 $M_\odot$ for the 5×10^4, 10^5 $M_\odot$ models, respectively). We note that these ISM masses are small compared to the stellar mass of the galaxy ($\sim 10^9$ $M_\odot$), and if the galaxy has a homogeneous distribution of nitrogen, that would rule out nitrogen enrichment by a single V/SMS. This issue could be resolved if the nitrogen has an inhomogeneous distribution (e.g. concentrated near the site of the V/SMS) or if multiple V/SMSs were present (as has been suggested by other authors: Cameron et al. [3], Charbonnel et al. [5]. This problem highlights one of the strengths of the V/SMS enrichment scenario, namely the large mass of nitrogen produced ($\sim$10–100 $M_\odot$). Other exotic enrichment scenarios (e.g. tidal disruption, peculiar supernovae yields) which do not produce as much nitrogen per event will have to have occurred many times in order for nitrogen to be observed at super-solar abundances.

The two high mass models may overproduce nitrogen relative to oxygen and veer out of the fiducial constraints on the left side of the grey region, the edge of which is not shown here (see Fig. 1 of Cameron et al. [3]). That being said, this depends heavily on the pre-SMS oxygen abundance in the ISM, a quantity which could feasibly span several orders of magnitude, and we argue that these two models naturally fall within the fiducial constraints, as their primary nucleosynthetic characteristic is nitrogen production. More generally, super-solar nitrogen abundances with roughly solar oxygen abundances can be achieved by supermassive stars with intermediate end times, and the inclusion of additional intermediate mass models would fill in the region between the high mass and low mass curves in Fig. 3.18.

3.6 Prospects Using Current and Future Observatories

The most promising avenue for detection of these events is the observation of the supernova itself. We will examine this in more detail in future work, but here we present basic considerations. We have shown that α process explosions are visible out to redshift 12 and beyond by JWST and other wider field telescopes. The metal rich explosions are even more energetic while also being closer in distance, due to the fact that the SMS is not metal free. Since the α process GRSN can be detected out to redshifts of 10–30 [14, 19], it seems likely that the metal rich GRSNe will be easier to detect. This will eventually allow the formation channels to be constrained (Sect. 2.4).

Recently, Moriya et al. [15] looked for GRSNe in early release JWST data. They searched for point sources which only occurred in the F444W and F356W bands. They did not find any of these point sources, and so constrained the GRSN rate to be less than $\sim 8 \times 10^7$ Mpc^3 yr^1. This rate is not yet low enough to test any of the scenarios outlined in the introduction, but if there continues to be no detection over the lifetime of JWST, then the constraint will become relevant. In comparison to the results discussed in this paper, our explosions appear in more than just the F444W and F356W bands. This is true both of the metal free explosions which are more luminous than the one discussed in [14] and the metal rich explosions, which would have occurred at a lower redshift, thus appearing in bluer bands. In the future, we plan to perform follow up studies similar to Moriya et al. [15], but searching for a wider variety of GRSNe.

The supernova, however, is by no means the only pathway towards identifying one of these explosions. GRSNe ejecta are thought to reach distances of several hundred kpc before falling back into the host halo and igniting a violent starburst [11, 29, 30]. One high redshift, luminous galaxy undergoing a starburst is GN-z11 (Sect. 3.18). This galaxy was recently observed by NIRCam to have a haze, although a definite association with the galaxy is not confirmed [26]. This haze could be evidence of a past merger or a supernova remnant. GN-z11 has also been observed by NIRSpec to have strong nitrogen lines [2] which have been tentatively interpreted as super-solar nitrogen [3]. If confirmed, both the haze and the super-solar nitrogen would be circumstantial evidence of a GRSN. Convincing evidence may be hard to nail down, but observations of Cl, K, Sc, or V would be a step in that direction.

Yoshii et al. [34] observed a high redshift quasar with [Mg/Fe] $= -1.11 \pm 0.12$. This observation is notable for its high iron abundance, and they suggest that it could be evidence of a pair instability supernova [27]. The metal rich GRSN, however, could also provide a plausible pathway towards explaining this abundance ratio, as the explosion both consumes magnesium and produce iron (Fig. 3.16). We also note that typical ISM metallicities at this redshift are much higher than the metallicity threshold for the SMS explosion [23]. Finally, in the galactic neighborhood, a population of nitrogen-enhanced stars has recently been observed [7], but these stars do not appear to have other signatures of the metal rich GRSNe.

References

1. Asplund M, Grevesse N, Sauval AJ, Scott P (2009) Annu Rev Astron Astrophys 47:481
2. Bunker AJ, Saxena A, Cameron AJ et al (2023). arXiv e-prints arXiv:2302.07256
3. Cameron AJ, Katz H, Rey MP, Saxena A (2023). arXiv e-prints arXiv:2302.10142
4. Chandrasekhar S (1964) Astrophys J 140:417
5. Charbonnel C, Schaerer D, Prantzos N et al (2023) Astron Astrophys 673:L7
6. Chen K-J, Heger A, Woosley S et al (2014) Astrophys J 790:162
7. Fernández-Trincado JG, Beers TC, Minniti D et al (2020) Astrophys J 903:L17
8. Fuller GM, Woosley SE, Weaver TA (1986) Astrophys J 307:675
9. Haemmerlé L (2021) Astron Astrophys 647:A83
10. Jiang L, Kashikawa N, Wang S et al (2021) Nat Astron 5:256
11. Johnson JL, Whalen DJ, Even W et al (2013) Astrophys J 775:107
12. Magg M, Nordlander T, Glover SCO et al (2020) Mon Not R Astron Soc 498:3703
13. Montero PJ, Janka H-T, Müller E (2012) Astrophys J 749:37
14. Moriya TJ, Chen K-J, Nakajima K, Tominaga N, Blinnikov SI (2021) Mon Not R Astron Soc 503:1206
15. Moriya TJ, Harikane Y, Inoue AK (2023) Mon Not R Astron Soc 526:2400
16. Nagele C, Umeda H (2023) Astrophys J 949:L16
17. Nagele C, Umeda H (2024) Phys Rev D 110:L061301
18. Nagele C, Umeda H, Takahashi K (2023) Mon Not R Astron Soc 523:1629
19. Nagele C, Umeda H, Takahashi K, Maeda K (2023) Mon Not R Astron Soc 520:L72
20. Nagele C, Umeda H, Takahashi K, Yoshida T, Sumiyoshi K (2020) Mon Not R Astron Soc 496:1224
21. Nagele C, Umeda H, Takahashi K, Yoshida T, Sumiyoshi K (2022) Mon Not R Astron Soc 517:1584
22. Oesch PA, Brammer G, van Dokkum PG et al (2016) Astrophys J 819:129
23. Pallottini A, Ferrara A, Gallerani S, Salvadori S, D'Odorico V (2014) Mon Not R Astron Soc 440:2498
24. Senchyna P, Plat A, Stark DP, Rudie GC (2023). arXiv e-prints arXiv:2303.04179
25. Shapiro SL, Teukolsky SA (1983) Black holes, white dwarfs, and neutron stars: the physics of compact objects
26. Tacchella S, Eisenstein DJ, Hainline K et al (2023). arXiv e-prints arXiv:2302.07234
27. Takahashi K, Yoshida T, Umeda H (2018) Astrophys J 857:111
28. Umeda H, Hosokawa T, Omukai K, Yoshida N (2016) Astrophys J 830:L34
29. Whalen DJ, Johnson JL, Smidt J et al (2013) Astrophys J 777:99
30. Whalen DJ, Johnson JL, Smidt J et al (2013b) Astrophys J 774:64
31. Woods TE, Heger A, Haemmerlé L (2020) Mon Not R Astron Soc 494:2236
32. Woods TE, Heger A, Whalen DJ, Haemmerlé L, Klessen RS (2017) Astrophys J Lett 42:L6
33. Woosley SE (2017) Astrophys J 836:244
34. Yoshii Y, Sameshima H, Tsujimoto T et al (2022) Astrophys J 937:61

Chapter 4
Discussion

Abstract We revisit the efforts of previous work on GR instability supernovae and contrast our approach and results with such studies. We highlight the importance of the GR stability analysis in determining the fate of the supermassive star, as stellar evolution codes cannot reliably find the GR radial instability. We then discuss shortcomings and sources of error associated with our approach, in HOSHI, the GR stability analysis, the hydrodynamics calculations, and the radiation hydrodynamics, respectively.

Keywords Supermassive stars · General relativistic radial instability · Thermonuclear supernovae

In this chapter, we turn to the question of the reliability and significance of our results. These are challenging questions to address. What we have done, in essence, is to study hypothetical types of explosions occurring in hypothetical objects, using three different numerical codes, not to mention the GR stability analysis, some of which may contain errors, and all of which suffer from a certain degree of uncertainty.

It is useful to review the understanding of GRSN before the work in this thesis. The study of GRSN can be said to have truly begun with Fuller et al. [5] and their study of metal rich explosions using KEPLER. This study was undoubtedly seminal, but suffered major drawbacks in the form of a small number of models, small nuclear network, and an incomplete treatment of GR. Next, we have a similar study by Montero et al. [7] which treated the problem in full relativity, but this came at the expense of an accurate progenitor structure and the inclusion of a reaction network.

Moving on to metal free GRSN, the first breakthrough came in Chen et al. [2], who discovered this phenomenon using KEPLER. However, this study only found a single, relatively unique, explosion, and suffered from a small nuclear network and an incomplete treatment of GR. Nagele et al. [9] remedied both of these concerns slightly, using a larger nuclear network and a fully relativistic hydrodynamics code for the dynamics, but we also only found a single unique model, and it was a different one than was found by Chen et al. [2]. Observables related to the Chen et al. [2] explosion were studied in several further papers [6, 8, 15, 16], but no systematic study had been performed.

© The Author(s), under exclusive license to Springer Nature Singapore Pte Ltd. 2024
C. Nagele, *General Relativistic Instability Supernovae*, Springer Theses,
https://doi.org/10.1007/978-981-96-0551-4_4

Let us compare this state of affairs to the work done in this thesis. We now understand that GRSNe are a much more general phenomenon than was previously thought, with a much wider mass range in the metal free and metal rich cases. We have performed a systematic analysis of the two most promising observables, nucleosynthetic yields and lightcurves, and compared them to JWST observations. The existence of GRSN as a general consequence of the existence of SMSs and the ability of current and future instruments to detect GRSN constitute a huge step forward in the study of SMSs. This fact is independent from the precise details of the explosions, and thus errors and uncertainties in our calculations would not fatally alter this conclusion. We will, of course, do our best to enumerate these errors and uncertainties, and devote the rest of this chapter to a discussion of the strengths and weaknesses of our methods.

HOSHI is a well developed stellar evolution code which has been used to study many aspects of massive star evolution [e.g. 11–13]. While the usual caveats, such as rotation and magnetic fields apply, we focus here on uncertainties related more specifically to SMSs. First is the question of initial conditions. As we have described, we initialize HOSHI models as very nearly isentropic $n = 3$ polytropes, which is the usual method for SMSs. However, in the case of extremely high accretion rates, such as in the galaxy merger scenario, this requires accretion to terminate on a short timescale. While the inflows on parsec scales are thought to terminate on these timescales, it is not clear that those inflows translate directly to accretion rates. On the other hand, for very low accretion rates onto the proto-SMS, nuclear burning will trigger before accretion has terminated and this case will be more similar to accreting SMS models. Next, we turn to the question of the consistency of HOSHI with GR. The HOSHI code includes the 1st order PN correction. In Nagele et al. [9], the pressure gradient was consistent with the general relativistic pressure gradient to order 10^{-2}. In Nagele et al. [10], we introduced the correction of the energy density (including, crucially, isotopic mass excess) to the relativistic density, which improved the consistency of the pressure gradient to order 10^{-5}. The 2nd and 3rd order TOV corrections would improve the consistency to 10^{-7} and 10^{-10} respectively, but we do not consider these additions necessary, as numerical error supersedes these values. Finally, we turn to the question of the $C^{12}(\alpha, \gamma)$ reaction rate. We adopt 1.5 times the rate of Caughlan and Fowler [1] (see discussion in [14]). As we have shown, the α process GRSN is heavily dependent on the amount of oxygen present at the time of the instability, so it is feasible that varying this rate could change the mass range of the explosion (either widen or narrow), just as is thought to be the case for PISNe [3, 4].

We now turn to the accuracy of the stability analysis. We have shown that the accuracy is very high for numerical polytropes, but that does not necessarily translate to the HOSHI models, which do not have as fine spatial resolution, and have irregular grid spacing. In addition, the timestep of the HOSHI simulations is high during helium burning ($\sim 10^9$ s) and the profiles are output once every five timesteps, meaning it is likely that the stability analysis does not find the instability as soon as it occurs. However, errors arising from poor spatial or temporal resolution are likely insignificant given the consistency of the GR stability analysis for similar HOSHI models. A more concerning discrepancy is that the hydrodynamics code occasionally finds stable models which are thought to be unstable by the stability analysis. However, this mostly happens with models which are still in hydrogen burning or are pre-main sequence. It is likely that there also exist models which are stable according to the stability analysis, but unstable according to the hydrodynamics code. Given the overall success of the stability analysis, however, as well as the computational cost of running multiple stellar profiles through the hydrodynamics code, we feel that our procedure optimizes our computational resources in locating the GR radial instability.

In Fig. 2.7, we attempted to quantify the error associated with mesh number, nuclear network size, and the rate of variation of the independent variables in the hydrodynamics code. Unlike the results of HOSHI, we do not expect the explosion to depend on the $C^{12}(\alpha, \gamma)$ reaction rate because this reaction is too slow to occur during the α process GRSN. Uncertainties in other reaction rates are a source of error in the CNO-rp explosions, but because no single reaction is crucial to the outcome, this is likely a minor issue. Chen et al. [2] showed that multidimensional effects can increase the explosion energy by a factor of 1.5, and due to this we likely underestimate the explosion energy of the GRSNe. Finally, we verified that the explosion is not effected by radiative transport.

There are many uncertainties when it comes to modeling shock breakout for supernova lightcurves. In this instance, however, we are aided by the similar calculation performed in Moriya et al. [8] using a different lightcurve code (STELLA). Our results for the explosions comparable to the one discussed in this study are remarkably similar, having in common maximum luminosity, temperature evolution, and the presence of a plateau due to hydrogen recombination. There are two major differences between our study and that of Moriya et al. [8], the first being the precise behavior of the lightcurve during the plateau period, with our results dimming in the redder bands, and the second being our identification of the pulsating models as considerably more luminous due to their large stellar radii in the late helium burning phase.

References

1. Caughlan GR, Fowler WA (1988) Atomic Data Nucl Data Tables 40:283
2. Chen K-J, Heger A, Woosley S et al (2014) Astrophys J 790:162
3. Farmer R, Renzo M, de Mink SE, Fishbach M, Justham S (2020) Astrophys J Lett 902:L36
4. Farmer R, Renzo M, de Mink SE, Marchant P, Justham S (2019) Astrophys J 887:53
5. Fuller GM, Woosley SE, Weaver TA (1986) Astrophys J 307:675
6. Johnson JL, Whalen DJ, Even W et al (2013) Astrophys J 775:107
7. Montero PJ, Janka H-T, Müller E (2012) Astrophys J 749:37
8. Moriya TJ, Chen K-J, Nakajima K, Tominaga N, Blinnikov SI (2021) Mon Not R Astron Soc 503:1206
9. Nagele C, Umeda H, Takahashi K, Yoshida T, Sumiyoshi K (2020) Mon Not R Astron Soc 496:1224

10. Nagele C, Umeda H, Takahashi K, Yoshida T, Sumiyoshi K (2022) Mon Not R Astron Soc 517:1584
11. Takahashi K, Umeda H, Yoshida T (2014) Astrophys J 794:40
12. Takahashi K, Yoshida T, Umeda H (2013) Astrophys J 771:28
13. Takahashi K, Yoshida T, Umeda H (2018) Astrophys J 857:111
14. Umeda H, Yoshida T, Takahashi K (2012) Progress Theor Exp Phys 2012:01A302
15. Whalen DJ, Johnson JL, Smidt J et al (2013) Astrophys J 777:99
16. Whalen DJ, Johnson JL, Smidt J et al (2013) Astrophys J 774:64

Chapter 5
Conclusion

Abstract We summarize and contextualize the research presented in this thesis.

Keywords Supermassive black hole problem · Early universe · General relativistic instability supernovae

In this chapter, we conclude with a summary of the work outlined in this thesis. Supermassive stars are hypothetical stars which are massive enough to experience the GR radial instability. They are studied due to their ability to form massive seed black holes in the early universe with an eye towards resolving the early universe SMBH problem, and due to the hot hydrogen burning temperatures which produce the correct nucleosynthetic yields to explain anomalous abundances in globular clusters. The formation of these SMSs is thought to be able to proceed via several pathways. First, if molecular hydrogen is destroyed in the early universe, then the resulting atomic gas has no method of cooling and this results in huge Jeans masses and SMS formation. A similar outcome can also be achieved if star formation is suppressed by supersonic baryon-dark matter streaming or turbulence due to cosmological inflows. In yet another scenario, the merger of two gas rich galaxies can cause gravitational torques which funnels the gas into a central rotationally supported disk. This disk collapses when it becomes massive enough, and this collapse can lead to SMS formation. Finally, globular clusters which accrete gas shrink in size while maintaining a fixed number of stars. This makes it more likely for stellar collisions to occur, and this can lead to a runaway process which produces a VMS or SMS.

It is all well and good to simulate the various formation scenarios on our supercomputers, but how can we actually verify whether or not SMSs really existed? Several possibilities have been considered, including direct detection of SMS photospheres, possibly aided by strong lensing, detecting SMS collapse via an ultra-long GRB or via gravitational waves, and finally detecting the copious neutrinos produced during SMS collapse. Although some of these approaches show promise, in this thesis we focus on the thermonuclear explosions of SMSs. Although they have shorter durations than the lifetimes of the SMSs, these events are extremely energetic and luminous, making them excellent candidates for detection by JWST, EUCLID and ROMAN, even in the very early universe. We also report that the nucleosynthetic yields from these events could be used to find evidence of a past GRSN.

In order to investigate GRSNe, we first construct evolutionary models of metal free and metal rich monolithic SMSs using the HOSHI codes. The models are initiated as pre-ZAMS, isentropic configurations which have similar structure to $n = 3$ polytropes. The models are then evolved to the point of the GR radial instability. For the metal free case, the SMSs of interest experience the instability in helium burning, while for the metal rich case, the SMSs of interest experience the instability in hydrogen burning. In the metal rich case, mass loss is non zero and can have a significant impact on the evolution.

In order to evaluate the GR radial instability, we numerically solve the pulsation equation for a normal mode decomposition of perturbations to the SMS. Once we have found the fundamental mode of the perturbation, the evaluation of stability is straightforward. When the SMS becomes unstable, we port the model to the hydrodynamics code with a nuclear network chosen for the type of explosion under consideration (61 isotopes for α process, 153 isotopes for CNO/rp process). The hydrodynamics code is then run in order to determine whether the model will explode or collapse to a black hole.

In the former case, we perform two analyses of potential observables. The first is to compute the nucleosynthetic yields for the ejecta. For α process explosions, this calculation is straightforward as the yields are dominated by α elements. However, for the CNO/rp process, we perform post-processing using at 514 isotope network which can capture the full rp process at low temperatures such as the ones considered here. The second form of analysis is to compute the lightcurve of the explosion using SNEC. We port the hydrodynamical models to SNEC at a time slightly before shock breakout, chosen so that the energy of the SNEC model matches the explosion energy of the hydrodynamical model. We then evolve SNEC past shock breakout, helium recombination, and hydrogen recombination, until the remnant becomes optically thin. During hydrogen recombination, the photosphere stalls and this 'plateau' phase of the lightcurve is a very strong candidate for detecting GRSN becomes of its long lifetime. Although the amount of variation of luminosity during this phase is not well understood, it may be able to be identified via color.

We then compare our results to JWST observations of super-solar nitrogen abundances in GN-z11. These abundances could be explained by hot CNO burning, on either explosive or evolutionary timescales. We compare the nitrogen abundances to yields from four models with metallicity $Z = 0.1 \, Z_\odot$ with varying masses, and conclude that the two higher mass models well fit the observations. Unlike other explanations for the super-solar nitrogen, the SMS explanation produces large amount of nitrogen (10–100 $M_\odot$) so that a single or a few SMSs could explain the abundances, rather than requires hundreds or thousands of rare events.

The work presented in this thesis has advanced the study of SMSs and GRSNe considerably, and will allow us to rigorously test for the presence of GRSNe in the near future. If GRSNe are found, then that will go a considerable way towards alleviating the early universe supermassive black hole problem. If they are not, then

5 Conclusion

it will be a strong piece of evidence that massive seeds were not common in the early universe, and other explanations will have to be examined more carefully. In either case, our understanding of the origin of early universe SMBHs will be advanced, and with it we will move one step closer to understanding the universe in its totality.

The manufacturer's authorised representative in the EU is Springer Nature Customer Service Centre GmbH, Europaplatz 3, 69115 Heidelberg, Germany. If you have any concerns regarding our products, please contact ProductSafety@springernature.com

Printed and bound by CPI Group (UK) Ltd, Croydon, CR0 4YY

26/03/2026

02078986-0002